The Focke-Wulf Fw 190

The Focke-Wulf Fw 190

William Green
and
Gordon Swanborough

DAVID & CHARLES
NEWTON ABBOT   LONDON   VANCOUVER

ISBN 0 7153 7084 7

© Pilot Press Ltd 1976

All rights reserved. No part of this publication may be reproduced, stored in a retrieval system, or transmitted, in any form or by any means, electronic, mechanical, photocopying, recording or otherwise, without the prior permission of David & Charles (Publishers) Limited

Set in 11 on 13pt Times
and printed in Great Britain
by Redwood Burn Limited
Trowbridge and Esher.
for David & Charles (Publishers) Limited
Brunel House Newton Abbot Devon

CONTENTS

Introduction	7
The Butcher-Bird of Bremen	9
Production for Service	27
The Search for Altitude	63
Metamorphosis	77
The Butcher-Bird in Combat	89
Technical Description	129
Development Batch, Prototypes, Production Variants	145

Introduction

More than three decades have now elapsed since the deep-throated growl of the BMW 801 air-cooled radial engine or the somewhat more cacophonous roar of the liquid-cooled inline Junkers Jumo 213 has been heard in the airframe of the Focke-Wulf Fw 190 or its derivative, the Ta 152; the World War II fighter progeny of Kurt Waldemar Tank. Their western contemporaries, the Spitfire, the Hurricane, the Mustang and the Thunderbolt, and even their comrade-in-arms, the Messerschmitt Bf 109, soldiered on well into the next decade, but Tank's progeny saw the end of their careers – apart from a singularly brief period with the French *Groupe de Chasse III/5 Normandie-Nieman* – with the termination of hostilities, less than four years after the first Tank fighter had initially fired its guns in anger.

The fact that the Fw 190, unlike its principal contemporaries, barely survived the end of World War II in Europe provided no measure of its success by comparison with the other principal fighters of the conflict. On the contrary, Kurt Tank's fighter was to be numbered among that élite group of truly great combat aircraft of all time; that distinguished assemblage of warplanes of all nations that together form military aviation's metaphorical hall of fame.

The Fw 190 – the *Würger*, or Butcher-Bird, as it was promptly if unofficially dubbed by its creator – was not an extraordinarily innovative design, but it embodied a number of innovatory features and if not representing a quantum leap in the state-of-the-art, the various characteristics that it offered combined to render it the most formidable fighter extant when first committed to combat. Like the Spitfire, it was an inspired design; unlike the Spitfire, it possessed no illustrious sires from experience with which its creator could draw. Yet, if undistinguished in pedigree, the Fw 190 was nonetheless a thoroughbred in every sense.

It took full advantage of the most advanced aerodynamic and structural techniques of the day, and shortcomings notwithstanding – and, indeed, what warplane ever designed could lay claim to total perfection – it had a greater impact on the air war fought in European skies than any fighter before or subsequently. If its introduction into the Luftwaffe inventory was from some aspects fortuitous, it was indeed fortunate for that service that it should become available at a time when the fortunes of the principal German fighter, the Messerschmitt Bf 109, appeared to be on the wane.

The success that attended the Fw 190 in its infancy was to be maintained during maturity. When the ascendancy that it had established in the skies above the channel coast diminished in the see-saw battle for equipment supremacy continuously waged by the combatants, the design of the fundamental components of the Focke-Wulf fighter facilitated a genealogical process embracing the application of progressively more powerful engines and armament, enabling it to retain a position in the forefront of its class until the end of hostilities.

The Fw 190 proved amenable to adaptation for a variety of rôles that were undoubtedly far from its creator's mind in 1938 when commencing its journey across the drawing boards in the Bremen design office; rôles such as close support in which the considerable success enjoyed was due in no small part to the fact that, from the outset of design, Tank had striven for greater structural integrity than was offered by any contemporary fighter.

In span of years, the career of the Focke-Wulf fighter may be said to have been brief, but if the Fw 190 did not enjoy service longevity, its achievements during the four years in which it flew and fought in European skies were dramatic enough.

Today, but a handful of the tens of thousands of Focke-Wulf fighters manufactured during the war years survive in various of the world's air museums; a few examples of one of the world's most outstanding fighters of all time that are guaranteed to set the adrenalin flowing of those of us who encountered the Fw 190 during those stirring years now so long past.

It is to this masterpiece of aeronautical design and engineering that this book is intended as a small tribute and in the preparation of which we had the invaluable assistance of W J A 'Tony' Wood in surveying the combat history of the Fw 190. We would also record our gratitude to Eberhard-Dietrich Weber whose meticulous notes on variants and sub-types have ensured a standard of accuracy that would otherwise have been unobtainable.

Finally, we would draw the reader's attention to the remarkably detailed 'cutaway' drawings contributed by John Weal and would acknowledge the expertise and eye-for-detail of Dennis I Punnett who has been responsible for the many other line drawings appearing in the pages that follow. To all those who were associated with the design, manufacture and operation of Kurt Tank's fighters and thus, inadvertently, played a rôle in this book, we would wish *Hals-und-Beinbruch*!

W G & G S

The Butcher-Bird of Bremen

Under an unseasonably overcast sky on 14 July 1971, the Federal German Republic's first post-war commercial airliner taxied quietly towards the end of Bremen Airport's 6,500-ft (2 000 m) main runway, the distinctive whine of its turbofan engines rising through the harmonic scale as pilots Nielsen and Bardill prepared to take their new bird into the air for the first time. For the previous three months this VFW 614 prototype had been undergoing final assembly and fitting-out in the VFW-Fokker company's main plant adjacent to the airport. Now the work-force, several thousand strong, had turned out to a man to watch their latest creation launched into its element, and to applaud. Once again, history was in the making; for this airfield, for this plant and even for some of these watchers, it was far from being the first time...

Even the time of year seemed to be right. Those among the watchers with a long enough memory could recall an earlier July, 34 years before, when the city of Bremen had witnessed the first take-off of another significant aircraft designed for airline use – the Focke-Wulf Fw 200 Condor. Just as the VFW 614 was now attracting the attention of all onlookers with the unique over-wing installation of its two engines, so had the Condor, in its day, stood out as the first European example of all-metal flush-riveted stressed-skin construction applied to a large four-engined cantilever low-wing monoplane. And then, two years later, had come the Fw 190: the Butcher-Bird of Bremen.

The Focke-Wulf Fw 190 V1 – the original 'butcher bird (*Würger*) of Bremen' – is seen in this photograph on an early test-flight. Unlike the production aircraft that were to follow, it was powered by a BMW 139 two-row radial engine, the cooling of which, through the small central orifice in the spinner, proved to be inadequate

The Fw 190 V1 nearing completion in the experimental shop at Focke-Wulf's Bremen factory – with an Fw 189 twin-boom reconnaissance aircraft just visible behind

From its inception the Condor had been intended for service with the German airline Deutsche Lufthansa, but in the event only sixteen examples were built in civil guise, and this aeroplane was destined to achieve greater fame, and notoriety, in the hands of pilots of the Luftwaffe, the air force of Germany's Third Reich that was already, in 1937, fast becoming the best equipped in Europe, if not the world. And it was the specific, anticipated, needs of that air force that were to lead to that other first flight at Bremen a mere two years after the Fw 200 had received its aerial baptism. Whereas in 1937 the airliner had been the subject of wide publicity, by the middle of 1939 the clouds of war were thickening fast and a security blanket had been thrown over the activities of Germany's armaments industry. Consequently, only those with a direct involvement were present to watch the first flight, on 1 June of that year, of what was to become the most noteworthy of all the varied aircraft types to emanate from the Focke-Wulf complex of factories in the Bremen area.

Like the Condor, the Fw 190 was the inspired product of a versatile and talented engineer, Dipl Ing Kurt Tank, who had joined Focke-Wulf in 1931. The company itself had been founded on 1 January 1924 as Focke-Wulf Flugzeugbau AG, its instigators being Heinrich Focke, George Wulf and Dr Werner Naumann. Focke and Wulf had collaborated in the construction of several primitive aircraft before World War I, and had resumed their joint activities in 1921 to produce the A 7 Storch two-seat monoplane. Working initially in a hangar at Bremen Airport shared with Deutsche Aero Lloyd and later occupying its own premises adjacent to the airport, the Focke-Wulf company produced, between 1924 and 1931, a series of light aircraft and small transports, the majority of which were monoplanes. The level of activity was sufficient to keep the company in business through some difficult times, but the designs were in no way remarkable and the name Focke-Wulf remained largely unknown outside Germany.

The event destined to change all that was the appointment to the company as its chief of design and flight testing of Dipl Ing Kurt Waldemar Tank, effective on 1 November 1931. Thirty-three years of age at the time of his appointment, Kurt Tank had been a design engineer with Rohrbach Metallflugzeugbau from 1924 to 1930, working on various of that company's flying boats and being primarily responsible for the Ro IX Rofix single-seat all-metal fighter monoplane. He had then spent 18 months working under Willy Messerschmitt as chief of the project bureau at the Bayerische Flugzeugwerke AG at Augsburg, but was forced to seek employment elsewhere when the latter filed a petition of bankruptcy on 1 June 1931.

Messerschmitt's loss was to prove to be Focke-Wulf's gain. At Bremen, Tank found a work force of 150 and work in various stages of completion on some half-dozen types of aircraft, among them the Cierva autogyro (under licence), three light transports, a light plane and a training biplane. The last mentioned was the A 44 (later Fw 44) Stieglitz, which in the next few years was to be instrumental in putting the name of Focke-Wulf firmly on the aviation map, being widely exported and produced under licence in five other countries. While overseeing the final development of this trainer, Tank embarked, in 1932, on the design of a Defence Fighter for which an official specification had been issued. This emerged in

Another view of the prototype under construction showing the tight-fitting cowling and the ducted fairing that rotated with the VDM propeller. A spinner was also fitted, inside the duct, prior to first flight

11

Above and below, left, the Fw 190 V1 during the initial taxying trials at Bremen, prior to the application of the registration letters. The cooling fan intended to be installed within the duct was not fitted at this stage

1933 as the Fw 56 Stösser, an attractive parasol-wing monoplane that was destined to win a fly-off against competing Arado and Heinkel designs in 1935, and became the subject of successive large-scale contracts that kept it in production until 1940, primarily for use by the *Jagdfliegerschulen* (Fighter Pilot Schools), in which it served throughout the war.

Between 1933 and 1939 Tank was responsible – as Technical Director of the company, which was reorganised in June 1936 as Focke-Wulf Flugzeugbau GmbH and became a subsidiary of AEG – for a series of aircraft that included, as well as the notable Fw 200 Condor, such unusual types as the Fw 187 single-seat twin-engined fighter, the Fw 62 catapult floatplane, the Fw 159 parasol monoplane fighter with retractable undercarriage, the Fw 57 twin-engined fighter-bomber and the Fw 189 twin-boom reconnaissance aircraft. Clearly, there was no lack of inventiveness in the design team at Bremen, and with the growing awareness of this fact in the *Reichsluftfahrtministerium* (RLM), plus the demonstrated excellence of the Stieglitz and Stösser, the Focke-Wulf company did not go short of official contracts for prototypes and production quantities. Expansion of the production facilities followed, new factories being occupied by Focke-Wulf in the Bremen suburbs of Hemlingen, Hastedt and Neuenland: eventually, they would be dispersed far to the east to be less vulnerable to RAF attacks, the flatlands of lower Saxony being among the most accessible to raiders from the UK.

By the time the Condor had been launched on its flight tests – Kurt Tank himself conducting the first flight – the Bremen company and its design and production teams were well placed to enjoy the confidence of the Air Ministry and its technical departments. This confidence was clearly demonstrated when Focke-Wulf was invited to consider the possible design of an advanced new fighter that could be ready, when the time came, to succeed the Messerschmitt Bf 109 – which was only then in the early stages of its service introduction. The launching of such a programme had not been given unanimous support either within the *Technische Amt* (Technical Office) of the RLM or the Luftwaffe itself, since the whole idea of such a *zweites Eisen im Feuer* (second iron in the fire) was considered by many to be quite premature.

For Kurt Tank and his team, on the other hand, the invitation was a golden opportunity to put into practice the lessons of the past few years, coupled with passionately held theories on optimum fighter design. A coherent design philosophy was regarded as

Above and below, the prototype Fw 190 V1 during the initial phase of test flights conducted by *Dipl-Ing Flugkapitän* Hans Sander. Bearing the registration D-OPZE, the aircraft is in a two-tone camouflage scheme, with light grey undersides

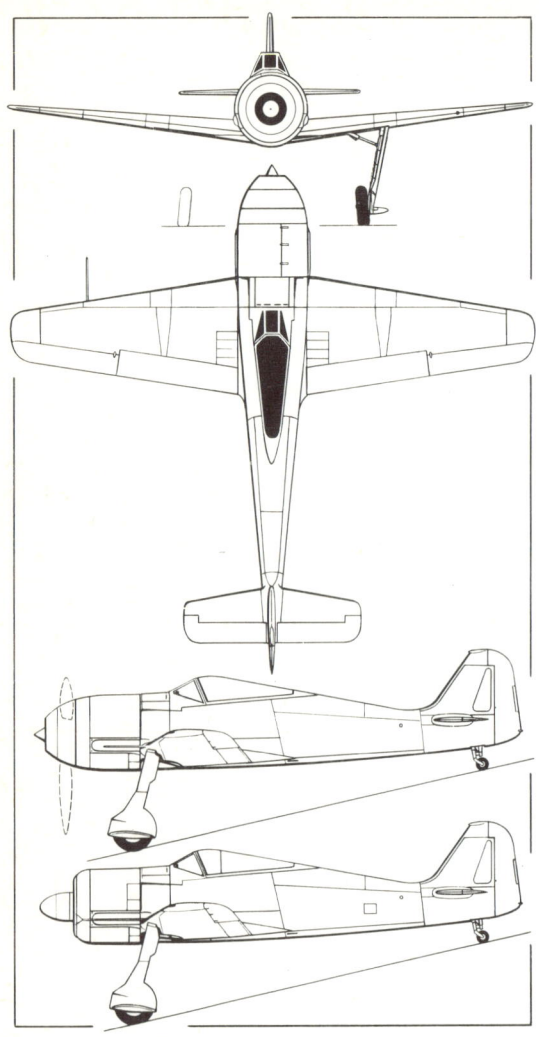

The three-view drawing depicts the Fw 190 V1 as it appeared for the first series of flight trials in 1939 and the lower side view shows the same aircraft after modification of the cowling

being paramount, with emphasis placed upon structural integrity, low airframe weight, simplicity of manufacture and field maintenance and, of course, high performance. A series of design proposals was submitted to the RLM by Focke-Wulf during 1938, incorporating Tank's ideas; most of these projects envisaged the use of a liquid-cooled in-line engine of the type favoured up to that time for the Luftwaffe's fighters, but a thorough analysis convinced Tank at this time that he could achieve the results he desired using an air-cooled radial engine.

Such a power plant was under development at the Bayerische Motoren Werke in Munich as the BMW 139 – a powerful 14-cylinder two-row radial that was, in terms of power demonstrated on the test bench, already some two years ahead of the in-line DB 601 and Jumo 211. Apart from this consideration, the BMW 139 was attractive to Tank because he foresaw that the massive fighter and bomber production programmes then under way in Germany were likely to monopolise the entire available output of the Daimler-Benz and Junkers engines. Moreover, he believed that a properly designed installation of a radial engine need not suffer excessive drag penalties and that it would prove less susceptible to battle damage than the liquid-cooled power plant.

That the *Technische Amt*, for long dedicated to the use of in-line engines, allowed itself to be swayed by these arguments was a tribute both to the forcefulness with which Kurt Tank argued them, and to the respect with which his views were now regarded. The fact that the programme was regarded as long-term, and scarcely more than an insurance policy against the possible failure of advanced developments of the Bf 109 to materialise in due course, no doubt also made it easier for the *Technische Amt* to give Focke-Wulf a contract, before the end of 1938, for three prototypes of the radial-engined fighter, to be designated Fw 190 in the RLM-approved series.

Almost from the start of its activities, the Focke-Wulf company had bestowed bird names upon its aircraft. This practice had begun with the A 17 Möwe (Gull) transport – the same name being used later for the 10-seat A 38 – and had continued with such types as the tail-first F 19 Ente (Duck) in which George Wulf met his death in 1927, the A 20 Habicht (Hawk) light transport, the S 24 Kiebitz (Lapwing), the A 32 Bussard (Buzzard) and A 33 Sperber (Sparrow) light transports, the A 43 Falke (Falcon) three-seat monoplane and the Fw 44 Stieglitz

(Goldfinch). Tank was happy to continue the custom with his Fw 56 Stösser (Bird of Prey), Fw 58 Weihe (Kite), Fw 187 Falke (Falcon), Fw 189 Eule (Owl) and Fw 200 Condor. What better name, then, for the Fw 190 than Würger (Shrike) – the ferocious butcher-bird that showed no quarter in attacking its prey! Time was to prove the name well chosen – yet once selected, it was virtually never again used in references to the new Focke-Wulf fighter, the less descriptive Fw 190 cipher becoming universally accepted instead.

Plans and Prototypes

Having received the blessing of the German Air Ministry, through the contract for three prototypes placed by the *Technische Amt*, Kurt Tank was able to proceed confidently with the task of translating his ideas for an advanced new fighter into hardware. He delegated general design responsibility to Oberingenieur R. Blaser, with Obering Mittelhuber in charge of the project office and Obering Kather looking after the design office. Whereas many earlier Focke-Wulf designs, and most that would follow, were designed under pressure to meet early delivery targets, there was relatively little urgency about the Fw 190 programme, at least in those last few months before the start of World War II, and if the pace in the Bremen project and design offices was not exactly leisurely, it did allow a conscious effort to be made to subordinate the hundred-and-one details of design to the concept of a coherent design philosophy.

The overall layout of the Fw 190 was conventional enough: there was little likelihood that a fighter conceived in the late

Above right and below, the Fw 190 V1 in 1940 after the ducted spinner had been replaced by a conventional constant-chord cowling. The 10-blade cooling fan is seen fitted, and performance measuring equipment is carried on a probe projecting from the port wing leading edge

15

'thirties would be other than a low-wing monoplane with a single tail unit and a retractable tailwheel-type undercarriage, and Kurt Tank's team was not yet ready to deviate from such well established lines – although it would not hesitate, within the next few years, to put forward some of the most advanced and even far-fetched proposals for fighter aircraft of any studied before the collapse of the Third Reich. To achieve the aim of producing an aircraft suitable for manufacture in widely dispersed factories and by sub-contractors possibly having little experience of modern aircraft construction, the overall lines were kept relatively severe, with straight taper on the wing leading and trailing edges. The pilot was located as far forward in the fuselage as possible for the best view over the wing, and a one-piece clear-view canopy was provided, this feature being well ahead of its time and affording the pilot a 360-degree field of vision round the aircraft.

Because the main fuselage structure was relatively light, the weight of equipment and engine had to be concentrated as close to the centre of gravity as possible, and this was achieved by mounting the BMW 139 engine on short bearers straight off the front fuselage bulkhead and front face of the wing spar. This, in turn, produced a relatively short fuselage, so that the ground angle had to be quite sharp to provide clearance for the large diameter propeller. The consequences of this were to be seen in the long stalky mainwheel legs of the Fw 190, but there was the offsetting advantage that these had to be attached almost at mid-span on

A line-up of Fw 190 test aircraft at the company's airfield, probably sometime in 1941. From left to right are an Fw 190A–0 with original wing, the Fw 190 V1, an Fw 190A–0 with the definitive wing, and two more small-wing A–0s, one of which is testing the under-fuselage drop tank

the wing in order to retract inwards, so producing an exceptionally wide undercarriage track with excellent stability when taxiing and for cross-wind landings.

With an overall diameter of about 50 in (129 cm) the BMW 139 engine was closely cowled, the cowling lines being continued aft to form a unity with the fuselage contours and forward to blend into a ducted spinner for the propeller, the area of the air intake in the spinner nose being further reduced by a small pointed propeller boss. To achieve adequate cooling of a two-row radial engine with an output of 1,550 hp at take-off through such a small orifice obviously posed problems, which it was proposed should be overcome by fitting a 10-bladed fan inside the cowling immediately ahead of the front row of cylinders. Driven by a gear train off the propeller reduction gearing to rotate at about three times the propeller speed, this fan would, it was believed, increase the air mass flow through the cowling sufficiently to keep the engine temperature within acceptable limits.

The span of the one-piece wing of the Fw 190 was a mere 31 ft 2½ in (9 515 m); by comparison the Bf 109B spanned 32 ft 4½ in (9,9 m) while the Spitfire and Hurricane were relatively gigantic with their respective spans of 36 ft 10 in (11 23 m) and 40 ft 0 in (12,2 m). Although the wing span was to grow before the Fw 190 entered production, Tank's fighter was to remain one of the smallest employed in large numbers during World War II, and its comparatively high wing loading and low power loading, as first designed, were among the features that were to make it such an excellent fighting machine, in addition to permitting its development to accept more powerful engines and heavier armament. The design of the engine installation, however, was such as to make difficult the location of arma-

The first prototype of the Fw 190 to fly with the BMW 801 engine was the V5, actually the third aircraft to join the flight test programme since construction of the third and fourth aircraft was abandoned. The Fw 190 V5 is seen here with the original small wing, showing the wing-root gun ports

ment in the forward fuselage, as favoured by other German designers; instead, provision was made for up to two machine guns to be fitted in each wing root, with synchronisation gear to permit them to fire through the airscrew disc.

The Fw 190 prototypes took shape in the experimental shop at Bremen early in 1939, in the form described in preceding paragraphs, and the first emerged in May to begin taxiing trials. Responsibility for flight testing was in the hands of Dipl-Ing Flugkapitän Hans Sander, the company's chief of flight test who, having declared himself satisfied with the aircraft, made the first take-off from the Neuenlander Feld on 1 June 1939. Finished in the standard Luftwaffe camouflage pattern with mid-green and green-black upper surfaces and a light grey underside, the prototype carried the swastika emblem in a white disc on a red band across the fin and rudder, but in lieu of national markings on the fuselage it bore the registration letters D–OPZE, these being repeated beneath the wing. Following RLM practice, the first prototype (*Werk-Nr* 0001) was designated the Fw 190 V1 (V for *Versuchs* or experimental, 1 for the first in sequence).

With no armament, the Fw 190 V1 weighed in at 6,103 lb (2 768 kg) and, right from the start, proved itself to be a 'pilot's aeroplane'. Development of the engine and its cooling fan was lagging behind airframe construction, however, and in order not to delay the start of flight testing, Sander had agreed to fly the V1 without the fan. As a result, extremely high temperatures were induced in the forward fuselage and the pilots who shared in the early trials with the aircraft were given an extremely uncomfortable time, with temperatures up to 55 deg C (131 deg F) being recorded in the cockpit. Additional discomfort was occasioned by the seepage of exhaust fumes into the cockpit, the canopy of which could not be opened in flight or during engine ground-running, and the wearing of an oxygen mask became mandatory.

Problems such as these could obviously be overcome, given time, but a greater source of worry to Tank and his team of engineers as they prepared the Fw 190 V1 for its first phase of flight testing was the fact that the BMW 139 engine was itself in serious trouble and the Bayerische Motoren Werke was anxious to abandon this unit completely in order to develop the more promising and somewhat larger BMW 801. The latter was a 14-cylinder unit expected to produce 1,600 hp or more; it had approximately the same diameter as the BMW 139, but it was considerably heavier and longer, so that although it would fit the cross section of the Fw 190's fuselage without any major problems, its use would require a significant redesign of the fighter for greater take-off weights and to keep the CG in the right place. Reluctant as the Focke-Wulf team must have been to undertake a major redesign so early in the life of their new fighter, it was clearly more sensible to do so than to try to put pressure on BMW to continue its efforts with an engine that might, in the end, prove unsatisfactory. Within a few days of the first flight of the Fw 190 V1, therefore, Tank agreed to proceed with this course, and the *Technische Amt* of the RLM granted BMW's request to end work on the BMW 139.

In order not to interrupt the initial flight testing of the Fw 190, a small unit was kept in being to back up the BMW 139 engines in the first two prototypes and the second of these, the Fw 190 V2 (*Werk-Nr* 0002), flew on 31 December 1939, with the cooling fan fitted and the same ducted spinner as used on the first. In full military finish, with the fuselage code letters FO+LZ, the Fw 190 V2 was the first to carry armament, comprising one 7.9-mm MG 17 and one 13-mm MG 131 in each wing root. Meanwhile, the Fw 190 V1, following an early series of handling trials by experienced Luftwaffe pilots at the *Erprobungsstelle* Rechlin during which it received the code letters FO+LY in place of the civil registration, had been returned to Bremen to have the cooling fan fitted. When it returned to flight status on 25 January 1940, it was also fitted with an orthodox spinner and NACA-type cowling, since wind tunnel tests – soon to be borne out in flight – indicated that the ducted spinner offered only a negligible reduction in drag but probably aggravated the engine cooling problem. The cooling fan was retained with the new spinner and, after a comparison of flight data for the Fw 190 V1 and V2, this became the standard for all future aircraft and was also adopted on the Fw 190 V2. In the early months of 1940 the latter was used for initial firing trials at the *Waffenprüfplatz* (weapons proving ground) at Tarnewitz and for further handling trials at Rechlin, but it crashed after completing about 50 hrs flying, when the engine crankshaft broke.

Early handling experience with the FW 190

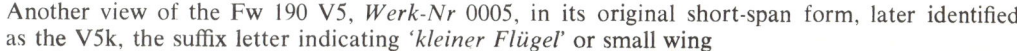

Another view of the Fw 190 V5, *Werk-Nr* 0005, in its original short-span form, later identified as the V5k, the suffix letter indicating '*kleiner Flügel*' or small wing

Above and below left, two views of the first of the development batch of Focke-Wulf Fw 190s used for development purposes. An Fw 190A–0 with BMW 801C–0 engine, this aircraft has the short-span wing and was known as the Fw 190 V6. It differed from later aircraft in that the rudder lacked the aerodynamic balance

prototypes revealed that the type had excellent manoeuvrability, good response to controls, low control forces, a high rate of roll and a high diving speed. Pilots were aware of some shortcomings, too, as would be expected in any prototype, but this did not prevent some impressive flying demonstrations being put on with the Fw 190, and when one of these was watched by Hermann Göring at Bremen at the beginning of 1940, he enthusiastically instructed Tank to turn out his new fighters *wie warme Semmel* (like hot rolls)!

Before the decision was made to redesign the Fw 190 to accept the BMW 801 engine, a fourth prototype had been ordered, and, together with the Fw 190 V3, this was completed structurally but was never flown because of the lack of BMW 139s. Eventually, the Fw 190 V3 was used as a source of spares for V1 and V2, while the completed V4 airframe was used for static tests. To replace these two aircraft, a fifth prototype was ordered during 1939, to be the first with BMW 801 engine, and work on a pre-production batch of forty was authorised to proceed, many of these being destined to emerge eventually as test aircraft.

To accommodate the BMW 801, which weighed some 350 lb (159 kg) more than the BMW 139, the Fw 190 had to be restressed throughout, and other changes were made. In particular, the wing was redesigned to have nil sweepback, whereas the original prototypes had a small degree of sweepback (measured on the line of greatest thickness from root to tip). The wing chord remained unchanged and there was only a fractional difference in span, at 31 ft 2 in (9 500 m), but the reduced angle of leading-edge taper resulted in the rake of the main undercarriage legs being changed when extended. Undercarriage retraction was changed from hydraulic to electric actuation, and small extensions of the wing leading-edge root were made to accommodate the wheel wells and provide more space for the wing-root gun installation. Doors to enclose the inner half of each wheel well were transferred from the main legs to hinge on the fuselage centreline, and the tailwheel was enlarged.

In company with these changes the cockpit was moved several inches aft relative to wing main spar, to help offset the greater weight forward of the new engine, and as a result Focke-Wulf's design team was able to offer a front-fuselage armament installation. At the same time the problem of high cockpit temperatures was alleviated, although other changes made to the cockpit were less felicitous: it was reduced in size, armour plate was added behind the pilot and the canopy was changed to incorporate a metal rear fairing, with the result that the all-round visibility was less satisfactory than hitherto, and taxiing became more difficult.

With wing loading and power loading both adversely affected by the increase in gross weight that accompanied the introduction of the BMW 801, it was inevitable that the fighter's performance should suffer, and this was quickly confirmed after the Fw 190

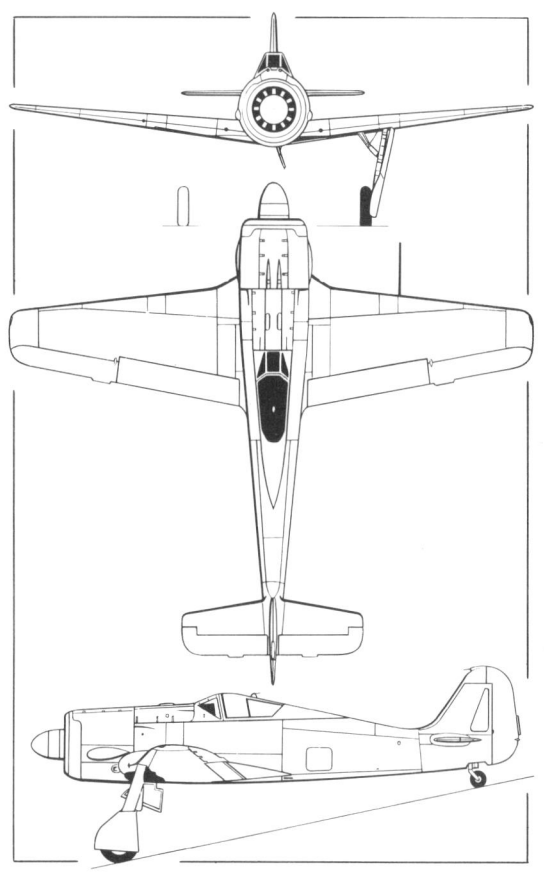

Above right, a three-view of the Focke-Wulf Fw 190A–0 in the initial production form, with the small-span wing that was fitted to only nine aircraft. Below, one of the initial Fw 190A–0s, showing the good accessibility of the BMW 801 engine through the hinged cowling panels

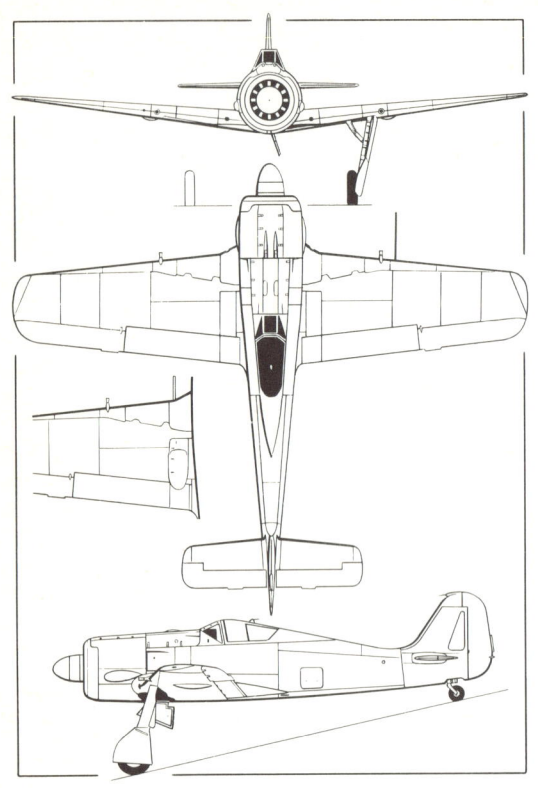

V5 (*Werk-Nr* 0005) entered flight test in April 1940. While the increase in take-off run was not critical, the reduced rate of climb and the impaired manoeuvrability demonstrated by this prototype *were* serious defects, to overcome which Obering Blaser's team proceeded to develop a larger wing for the Fw 190. This was achieved simply by extending the wing at each tip, to produce an overall span of 34 ft 0¾ in (10 383 m) and an area of 196.98 sq ft (18,3 m²). When this wing was fitted to the Fw 190 V5 later in 1940 (together with a larger tailplane), it was found that the earlier excellent handling characteristics had been restored, and the rate of climb was significantly improved, for the loss of 6 mph (9,6 km/h) in top speed. For reference purposes, the prototype in its two forms was designated Fw 190 V5k (for *Kleiner*, small wing) and Fw 190 V5g (for *grosser*, large wing).

The new wing was introduced as the production standard as rapidly as possible, but

Above left, a three view of the Focke-Wulf Fw 190A–0 with the definitive wing, this drawing also being representative of the Fw 190A–1. The inset detail shows the Fw 190A–2 wing-root installation of the 20-mm MG 151 cannon in place of the 7.9-mm MG 17 used earlier. Below, a short-span Fw 190A–0, *Werk-Nr* 0022, with trial installation of the fuselage drop tank

could not be phased in quickly enough to be fitted to the first of the batch of forty pre-production aircraft which Focke-Wulf had in production, and nine of the latter were completed with the original short span 'K' wing. This entire batch had been ordered as the Fw 190A–0, but the first example (*Werk-Nr* 0006) was delivered as the V6, completed at the end of 1940 and being similar in most respects to the V5k. The next airframe (*Werk-Nr* 0007) was set aside for static tests, so the first Fw 190A–0 to appear as such was *Werk-Nr* 0008; the batch continued to *Werk-Nr* 00035, for a total of twenty-eight; the final ten aircraft were reassigned for development of the later Fw 190B, C and D production series, being joined in this task by a further fifteen test aicraft ordered subsequently, the *Werk-Nummern* of which were assigned consecutively (ie, *Werk-Nr* 0046 to 0060 inclusive).

Many of the twenty-eight Fw 190A–0 airframes were also assigned special test roles, especially in respect of different armament installations. The basic armament originally intended for the A–0 comprised four 7,9-mm Rheinmetall Borsig MG 17 machine guns, two in the upper decking of the forward fuselage and two in the wing roots, all four firing through the propeller disc with the benefit of electric synchronisation gear. This armament was fitted in the Fw 190 V6; subsequently provision was made in this aircraft for the installation of a 20-mm MG FF cannon in each wing, just outboard of the main undercarriage attachment point, with 55 rpg, and the designation was then changed to Fw 190A–0/U1.*
Another early armament variation was to introduce provision beneath each wing –

---

* This was the first application of an *Umbau* number to the Fw 190; many more were to follow. *Umbau* meant 'modified' or 'rebuilt' and applied to one-off variants or small pre-production quantities used to prove new equipment or armament installations. Some later 'U' suffixes are thought to have indicated *Umrust-Bausatz* or factory conversion of larger quantities of aircraft to have a new armament or equipment standard, these being comparable with the 'R' suffix, indicating *Rüstsatz*, applied to conversion kits installed in the field.

The Fw 190A–0 *Werk-Nr* 0025, with long-span wing, was one of several aircraft from the development batch that were used for an early service evaluation by a detachment of personnel from II/JG 26 based at Le Bourget. The trial was somewhat catastrophic, many teething problems being encountered. This aircraft was later modified to Fw 190A–0/U13 standard

Above, one of the final group of Fw 190A–0s used for service trials; MG FF cannon are fitted in the outer wing gun bays. Below, the same aircraft is seen in company with two others of the same batch, posing for propaganda photographs taken by the Focke-Wulf photographer

and in due course beneath the fuselage also – for one 250-kg (551-lb) bomb or one 66 Imp gal (300 litre) fuel tank to be carried beneath each wing. This installation was developed on two of the trials aircraft under the designation Fw 190A–0/U4 (*Werk-Nummern* 0022 and 0023); later, the first of these was used for some unsuccessful ejection-seat trials and the latter had a trial installation of FuG 16Z communications transmitter/receiver while under test at Rechlin. The Fw 190A–0/U3 (*Werk-Nr* 0021) was the first to be fitted with FuG 25 IFF (identification, friend from foe) equipment; the Fw 190A–0/U5 (*Werk-Nr* 0018) had 20-mm MG 151 cannon in the wing roots in place of MG 17s, and the Fw 190A–0/U10 (*Werk-Nr* 0030) was similar

Above, another study of one of the Fw 190A–0s at the service evaluation stage. Like the aircraft opposite, it has 20-mm MG FF cannons in the outer wing positions, supplementing the smaller calibre MG 17 weapons in the wing roots

Above, the Fw 190A–0 *Werk-Nr* 0022 ended its useful life as a test-bed for early ejection-seat trials. Below left, another of the early Fw 190A–0 development aircraft used for service evaluation trials was *Werk-Nr* 0027

but had MG FF/Gs in the outer wing bays, these guns making use of linked cartridges.

Other early variations concerned the engine: seven of the A–0s (the sub-series U2, U3 and U4) were powered, like the prototypes, with the pre-production BMW 801C–0, but most had the early production BMW 801C–1, which had a similar rating and was first tested in the Fw 190A–0/U11 (*Werk-Nr* 0015), as well as being used in the Fw 190A–0/U5 and U10. The cooling fan of this engine was a 12-bladed unit, rather than the 10-bladed fan as designed originally for use on the BMW 139, and geared to operate at 3·17 times airscrew speed. Both these early engine versions were rated at 1,600 hp, but BMW was already working on a more powerful version, the 801D rated at 1,700 hp, and a trial installation of this was made in the Fw 190A–0 (*Werk-Nr* 0031) as the A–0/U12, while four others (*Werk-Nummern* 0025–28) because A–0/U13s as engine test-beds with this same engine; one other installation of the BMW 801D was made, in the Fw 190A–0 *Werk-Nr* 0014, which was in other respects a U2 sub-variant and then became known as the Fw 190A–0/U2/13 with the new engine. In the case of the A0/U12, a nitrous-oxide (GM–1) system of power boosting was also incorporated for test purposes.

Production for Service

In addition to their rôle in developing engine, armament and other equipment installations, the Fw 190A–0 series of aircraft also played a vital part in the service introduction of the 'butcher-bird of Bremen', six examples being assigned as early as March 1941 to an initial service test unit. This unit was the *Erprobungsstaffel* 190, formed with personnel detached from the II *Gruppe des Jagdgeschwaders 26 Schlageter*, which had already been earmarked as the first operational unit to receive Tank's new fighter. Early service experience at Rechlin and later at Le Bourget gave scant cause for enthusiasm (see p 91), and at one point in 1941 the entire programme appeared to be in jeopardy. Only the most strenuous efforts of Focke-Wulf and BMW engineers to overcome the many teething troubles associated with the power plant installation saved the day and ensured continuation of the production batches that had meanwhile been put in hand.

Production had been launched, in fact, some months before this crisis point was reached in the Fw 190's career, when the

Another photograph of the Fw 190A–0 *Werk-Nr* 0022 (see opposite page), while being used for *Jabo* tests in the fighter-bomber role, as the A–0/U4. It had also previously flown with a drop tank under the fuselage, as shown on page 22

Above and below left, the Fw 190A-1/U1 *Werk-Nr* 098 with a full load of SC 50 bombs under the wings and fuselage, for operation in the *Jagdbomber* (fighter-bomber) rôle. The SC 50 bomb was a 110-lb (50-kg) weapon

Technische Amt placed a contract with Focke-Wulf for 102 of the Fw 190A-1 variant, to be built at the new dispersal factory that the company was setting up at Marienburg. At about the same time the AGO and Arado companies were instructed to prepare for sub-contract manufacture of the Fw 190 at Oschersleben and Warnemünde respectively.

Deliveries of the production aircraft from Marienburg began in mid-1941, and the rate built up rapidly, all 102 having been handed over to the Luftwaffe by October. The first aircraft (*Werk-Nr* 001* – not to be confused with 0001, which was the Fw 190 V1) was designated the Fw 190 V7 to serve as the prototype of the Fw 190A-1 series, the general standard of which was similar to that of the A-0, with the BMW 801C-1 engine. Among the changes were the use of a cartridge jettison system for the cockpit canopy, which in early trials had proved reluctant to part company with the aircraft at speeds above 250 mph (402 km/h), and heavier toggle latches to secure the engine cowling in position.

Armament of the Fw 190A-1 comprised the basic four MG 17s – two in the front fuselage decking and two in the wing roots – plus the two MG FF cannon farther outboard in the wings, and provision for carry-

---

* The full *Werk-Nr,* in fact, was 190.0110.001, in which the group '0110' indicated the first sub-series of the first production variant (ie, the A-1). Thus, the first Fw 190A-2 had the *Werk-Nr* 190.0120.001, the first Fw 190A-3 was 190.0130.001 and so on. Fw 190B production would start at 190.0210, the Fw 190C at 190.0310, the Fw 190D at 190.0410, the Fw 190E at 190.0510, the Fw 190F at 190.0610 and the Fw 190G at 190.0710.

ing a bomb or fuel tank externally beneath the fuselage. A Revi C/12D reflector gunsight was fitted, and fire selection equipment permitted any pair of guns to be fired separately, or any combination of pairs to be fired together. Standard equipment included the FuG 7 radio, frequently supplemented by an FuG 25 set, and armour protection for the pilot's back and head. The fully equipped weight had by now risen to 8,488 lb (3 850 kg), so performance was beginning to suffer, particularly when a full external load was being carried. As soon as possible, therefore, the more powerful BMW 801D was introduced, this being rated in its D-2 version at 1,705 hp at sea level and at 1,500 hp for the initial climb. Post-production modification of some Fw 190A-1s to have this engine was indicated by the designation Fw 190A-1/U1, while its introduction on the production line changed the designation to Fw 190A-3, as noted below. Before this stage was reached, however, the Arado and AGO production lines began delivering the Fw 190A-2, in August and October 1941 respectively, and in the latter month the A-2 also succeeded the A-1 on Focke-Wulf's own production line.

The original armament planned for the Fw 190, comprising only four 7.9-mm MG 17s, was clearly inadequate for the kind of aerial warfare that was being waged over Europe by 1941, and this had already been realised by the Focke-Wulf team. The addition of the two 20-mm MG FF cannon in the outer wing almost from the start of production helped, and a further stage in armament development was reached when suitable interrupter gear was produced to permit the 20-mm MG 151 weapon to be installed in the wing roots of the Fw 190. Tested on the Fw 190A-0/U5, this installation became standard on the Fw 190A-2, the specific prototype of which was designated Fw 190 V14, this being the first production model. Apart from the introduction of the MG 151s, which could be distinguished by small fairings over the breech blocks on the wing upper surfaces, the Fw 190A-2 variant was similar to the A-1 with the BMW 801C-2 engine. Production totalled 426, of which Focke-Wulf built 118, Arado, 203 and AGO, 105. One example was

Factory-fresh Fw 190A-3s awaiting delivery. The light green mottle finish was characteristic of almost all early production Fw 190s

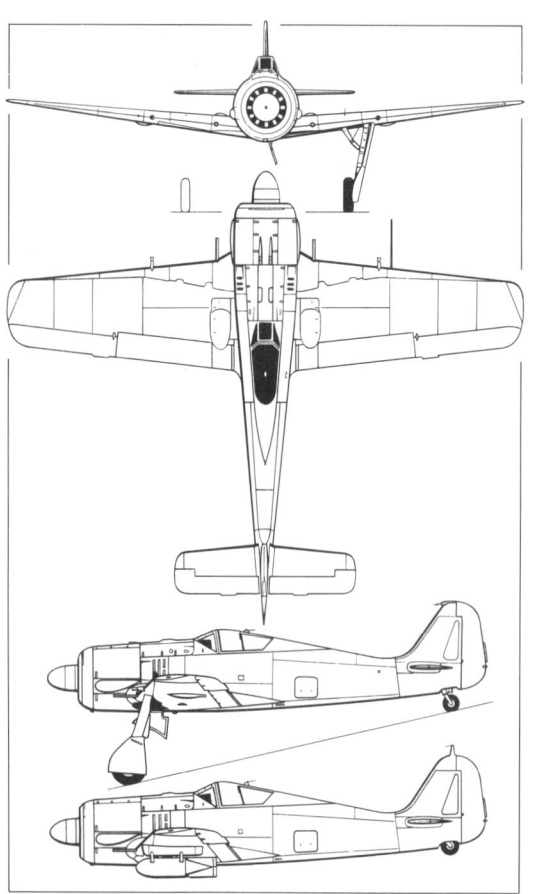

modified to carry a Patin PKS directional control system (in effect, an auto-pilot controlling only the rudder control circuit) as the Fw 190A–2/U1 (*Werk-Nr* 315).

In addition to its use as an interceptor, the Fw 190 quickly gained a reputation as a fighter-bomber (*Jagdbomber*) and an extended-range fighter-bomber (*Jagdbomber mit vergrösserter Reichweite*), usually referred to as the *Jabo* and *Jabo-Rei* versions. This followed the trials made with the two Fw 190A–0/U4s and, subsequently, the A–1/U1, and involved the fitting of an ETC 501 rack beneath the fuselage to carry one SC 250 or one SC 500 bomb, respectively weighing 250 kg (551 lb) and 500 kg (1,102 lb), and racks beneath the wings to carry four SC 50 bombs of 50 kg (110 lb) each. In place of wing bombs, a 300 litre (66 Imp gal) drop tank could be carried under each wing. When the ETC 501 rack was fitted, the wheel-well half doors were deleted from the fuselage.

The Fw 190A–3, which began to leave the production lines in 1942, was fully equipped as a fighter-bomber, and differed

Above left, a three-view drawing of the Focke-Wulf Fw 190A–3 with an additional side-view of the Fw 190A–4/U1, showing the revised aerial attachment atop the fin and the underwing bomb racks. Below, an early production Fw 190A–1 with MG FFs in the outer wing bays

A Focke-Wulf Fw 190A-3 production model showing the MG 151 cannon in the inner wing bays, but with no guns in the outer bays, indicating that the aircraft was in use for trial purposes

from the A-2 primarily in that the uprated BMW 801D-2 engine was fitted, with benefit to the overall performance. No prototype of the Fw 190A-3 series appears to have been designated as such, the first installations of the 'D' engine having been made in various of the A-0 series aicraft, as noted earlier. Production was undertaken by all three concerns already involved in the Fw 190 programme, and also by the Gerhard Fieseler Werke at Kassel, which produced its first A-3 model in May 1942. Production of the Fw 190 had built up extremely rapidly, the Third Reich being as yet relatively unaffected by the Allied air attacks that would, within the next two years, take their toll of virtually every production plan prepared by the *Technische Amt*. By the end of 1941, 224 Fw 190s had been accepted by the Luftwaffe; during 1942 a total of 1,878 was delivered, with a production peak of 194 attained in July.

Two examples of the Fw 190A-3 were modified by Focke-Wulf for special test purposes: *Werk-Nr* 270 became the

Fw 190A-3 Specification

Power Plant: One BMW 801D-2 14-cylinder radial air-cooled engine rated at 1,700 hp for take-off and 1,440 hp at 18,700 ft (5 700 m).

Performance: Max speed, 312 mph (502 km/h) at 19,685 ft (6 000 m) and with one-minute override boost, 418 mph (673 km/h) at 21,000 ft (6 400 m); initial rate of climb, 2,830 ft/min (14,4 m/sec); time to climb to 26,250 ft (8 000 m), 12 min; service ceiling, 34,775 ft (10 600 m); max range, 497 mls (800 km).

Weights: Empty, 6,393 lb (2 900 kg); empty equipped, 7,110 lb (3 225 kg); normal loaded, 8,770 lb (3 980 kg).

Dimensions: Span, 34 ft $0\frac{3}{4}$ in (10,383 m); length, 28 ft $10\frac{1}{2}$ in (8,798 m); height, (over airscrew), 12 ft $11\frac{1}{2}$ in (3,95 m); wing area, 196,98 sq ft (18,3 m²).

Armament: Two 7.9-mm MG 17 machine guns in fuselage with 1,000 rpg; two 20-mm MG 151/20 cannon in wing roots with 200 rpg and two 20-mm MG FF cannon in wings with 55 rpg.

Above and below, to permit the Fw 190 to operate in North Africa, tropical filters were developed for the carburettor air intakes; a trial installation is illustrated on this Fw 190A-3, with the intakes outside the cowling instead of being inside

A-3/U1 with an experimental installation of the BMW 801D in a lengthened mounting, eventually adopted by the Fw 190A-5 and subsequent production versions; and Werk-Nr 386 was tested in July 1943 with an experimental installation of the RZ 65 rocket projectiles as the Fw 190A-3/U2. A third trials aircraft, the Fw 190A-3/U3 (Werk-Nr 300), was the first example of the Fw 190 to carry reconnaissance cameras, comprising one RB 50/30 and one RB 75/30 in the fuselage behind the pilot. A small fairing was added under the fuselage to protect the camera lenses and the success of this concept led to the modification of twelve more examples in October–November 1942, these going into service as the Fw 190A-3/U4, the camera installation in this case comprising two RB 12,5/7×9 in the fuselage and one Robot camera in the port wing. In September 1942, three special lightweight models of the fighter were produced, as the Fw 190A-3/U-7 (Werk-Nummern 528, 530, 531), having BMW 801C engines and the fuselage guns removed. They were intended for high altitude use, but nothing is known of their operational deployment. The designation Fw 190A-3tp was used for three examples (Werk-Nummern 511, 514 and 515) employed by Focke-Wulf to develop tropical filters for the BMW 801D installation.

A batch of seventy-two Fw 190A-3s was built for Turkey under the terms of an agreement concluded between the respective governments in 1941. Designated Fw 190Aa-3, they were delivered, according to Focke-Wulf records, from October 1942 to May 1943 under the code-name 'Hamburg', and they entered service with the 3rd and 5th Squadrons of the Turkish Air Force's 5th Air Regiment at Bursa. Powered by the BMW 801D-2 engine, they had an armament of four MG 17s and two MG FFs.

An Fw 190A-3tp, *Werk-Nr* 511, with tropical filters incorporated in the carburettor air intake, and an ETC 501 rack beneath the fuselage, as used later on the Fw 190A-4/U3 *Jabo*

They were destined to outlive all other examples of the classic fighter from Bremen, save a few preserved as museum pieces, for they remained in service in Turkey until 1948.

The Fw 190A-4 followed the A-3 on the production lines from July 1942 onwards, production of the latter totalling 509 (including the versions described in previous paragraphs). The A-4 sub-series consolidated various production-line improvements that had been introduced in earlier models, but the most significant innovation was the use of FuG 16Z radio in place of FuG 7a, this

An Fw 190A-3, *Werk-Nr* 447, photographed on test in June 1942 with four 110-lb (50-kg) SC 50 bombs carried on an adapter on an ETC 250 rack beneath the fuselage

33

Appearance of the Fw 190 in combat gave *Luftwaffe* pilots a significant advantage over their RAF opponents. British assessment of the threat posed by the Fw 190 was facilitated by the inadvertent landing, in May 1943, of this Fw 190A–3 at RAF Pembrey. Photographs of this aircraft in RAF markings appear on subsequent pages

being distinguished externally by the addition of a small aerial pylon at the tip of the fin. As in the earlier models, FuG 25a IFF was usually carried in addition, with a ventral 'stick' antenna. The first production example (*Werk-Nr* 581) served as a prototype, designated Fw 190 V24, and a few aircraft fitted with the BMW 801C–2 engine during routine servicing, to overcome temporary shortages of the later engine, were designated A–4/U1. Thirty Fw 190A–4/U3s delivered in September and October 1943 were classified as *Schlachtflugzeug* (literally, battle aircraft) for use as tactical fighter-bombers in support of *Panzer* formations on the ground. They carried additional armour protection for the engine and pilot, standard built-in armament, and either one SC 250 bomb or one 300 litre (66 Imp gal) fuel tank under the fuselage. The Fw 190A–4/U8, two examples of which were built (*Werk-Nummern* 669 and 670), was the first *Jabo-Rei* variant, able to carry both a 500 kg (1,100-lb) bomb under the fuselage on an ETC 501 rack, plus two underwing drop tanks on Ju 87-type VTr racks, by virtue of having the fuselage guns and outer wing guns removed; earlier *Jabo* versions could not carry bombs *and* tanks simultaneously. The designation Fw 190A–4/U4 was reserved for a reconnaissance-fighter similar to the A–3/U4, but it is believed that no examples were converted, a version of the later Fw 190A–5 being used instead. Production of the Fw 190A–4 series was completed early in 1943 and totalled 894.

Installation of special FuG 16Z–E communications radio in some Fw 190s allowed them to be used as lead-fighters (*Leit-Jager*), the first such modifications being made in the Fw 190A–4/R1 variant. Another modification arising out of operational necessity in air combat over Germany produced the R6 *Rüstsatz*, also first applied to the Fw 190A–4, and comprising the installation of a Wfr Gr 21 rocket launcher beneath each wing. This weapon, firing 21-cm mortar shells, was used to break up tight formations of USAAF bombers, and was first

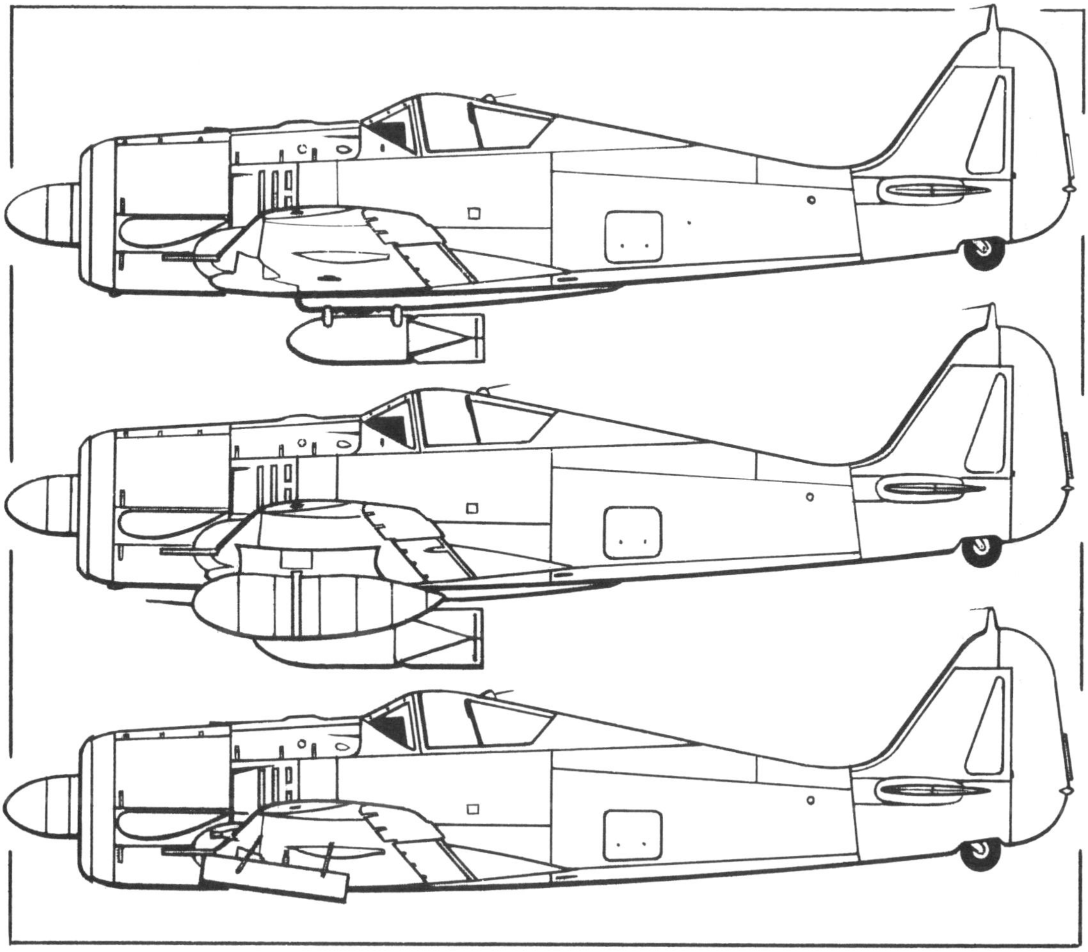

Above, three variants of the Fw 190A–4 are depicted in these side views of, top to bottom, the A–4/U3 *Jabo* with fuselage bomb-rack, the A–4/U8 *Jabo-Rei* with fuselage bomb-rack and wing drop tanks, and the A–4/R6 *Pulk-Zerstörer* with Wfr Gr 21 mortars under the wings. Below, a line-up of Fw 190A–4/U8 *Jabo-Rei* (fighter-bombers) with long-range drop tanks

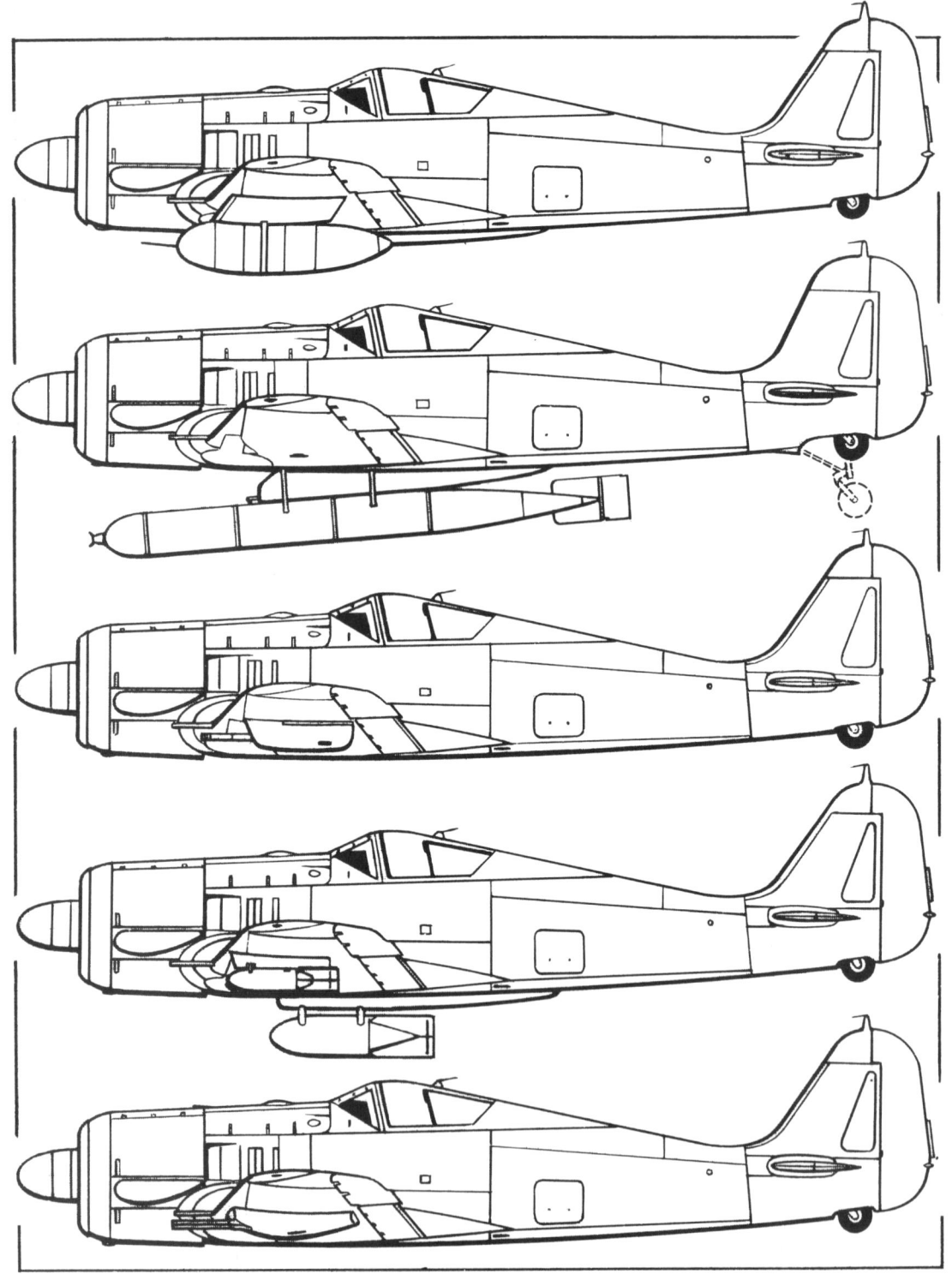

Further Fw 190A production variants depicted by these side views, top to bottom, are the A–5/U8 *Jabo-Rei* with bomb-rack and drop tanks; the A–5/U14 torpedo-fighter with LTF 5b torpedo on a modified ETC 502 rack, enlarged fin and lengthened tail wheel; A–5/U16, a single example with a 30-mm MK 108 cannon beneath each wing; A–5/U17 *Schlachtflugzeug* development aircraft with wing and fuselage bomb-racks, and A–6/R1 *Zerstörer* with a pair of MG 151 cannon beneath each wing used – with considerable success – over Schweinfurt on 14 October 1943, allowing the conventionally-armed Fw 190As to wreak considerable havoc among the B–17 and B–24 formations as soon as they had been dispersed. Whereas the *Umbau* modifications had tended to be numbered in sequence for each Fw 190A sub-type, the *Rüstsätze* were often applied to more than one sub-type, and thus the same R sequence number indicated the same modification for each sub-type; thus R1 lead-fighter versions of both the Fw 190A–5 and Fw 190B–1 were designated in due course (although the latter did not materialise) and Fw 190A–5/R6 mortar-armed fighters were operational alongside the Fw 190A–4/R6 versions from October 1943 onwards, with similarly armed Fw 190A–6/R6 and A–7/R6 versions appearing later.

The Fw 190A–5 had appeared late in 1942, and was in effect an A–4 with the new engine installation that had been developed

Above and below, an Fw 190A–4/R6 with armoured cowling ring and Wfr Gr 21 mortars beneath each wing. This was a *Pulk-Zerstörer* variant developed in 1943 to combat large formations of USAAF bombers and the use of 21-cm (9-in) mortar shells in this way achieved considerable success

A Focke-Wulf Fw 190A–5/U8 *Jabo-Rei* fighter-bomber with a 1,100-lb (500-kg) bomb under the fuselage and location of the underwing tank pylons clearly visible, although the pylons themselves are not fitted. In the background, extreme right, can be seen the Fw 190 V30 '*Kanguruh*' with ventral turbo-supercharger

on the A–3/U1; this increased the overall length by 15,25 cm (5¾ in) and compensated for the growing weight of equipment in the fuselage, which had been having an adverse effect on the handling characteristics. Production of the Fw 190A–5 by the four assembly centres totalled 723, and a large number of conversions was designated, especially in connection with new armament installations. The Fw 190A–5/U1, U3, U4 and U8 were respectively the same as the similarly designated A–4 versions; the U–3 *Schlachtflugzeug* had provision to carry four 50-kg (110-lb) SC 50 bombs on the fuselage rack by means of an ER 4 adaptor and was sometimes equipped for tropical operation, indicated by a 'tp' suffix on the designation. The Fw 190A–5/U4, also tropicalised, was an *Aufklarer* H (short-range reconnaissance) variant projected in place of the Fw 190A–4/U4, and the Fw 190A–5/U8 *Jabo-Rei* had the same armament mix as the A–4/U8.

In common with most other single-seaters of its day, the Fw 190 was a difficult aeroplane to use operationally at night because of the glare from the exhaust stubs, which were always in the pilot's line of vision and reduced his ability to see more distant targets. During 1943, Focke-Wulf conducted tests with three flame-damping systems, including shields beneath the stubs to reduce visibility of the aircraft to ground observers. After tests on a factory aircraft, two service Fw 190A–5s were fitted with the *Ofenschirm* (fire-screen) and *Kohlenkasten* (coalbox) screens developed by the Klatte company and were used for night flights and simulated night operations by Luftwaffe pilots at Rechlin. Five more examples were then modified, taking the designation Fw 190A–5/U2 (out of sequence, but an *Umbau* number not then applicable for other modifications), and saw limited service use in the ground attack rôle, the armament standard being that of the Fw 190A–5/U8 *Jabo-Rei*.

The designations Fw 190A–5/U9 to Fw 190A–5/U12 applied to a series of trials aircraft with modified armament, directed towards improving both the ground-attack

Above and right, two views of an Fw 190A-5/U8, *Werk-Nr* 825, with Tr Ju 87 fairings under the wings to carry 300-1 (66-Imp gal) drop tanks. Below, Fw 190 wing and fuselage loads, shown diagramatically, included: A, four SC 50 bombs on an ER-4 adapter; B, one SC 250 bomb; C, one SC 500 bomb; D, one SB 1000 bomb; E, one 300-1 (66-Imp gal) drop tank; F, two SC 250 bombs; G, four SC 50 bombs; H, two 300-1 (66-Imp gal) drop tanks; I, four MG 151s in the R1 modification for Fw 190-6, A-7 and A-8; J, two MK 108s in the single A-5/U16 *Werk-Nr* 1346; K, two MK 103s in the R3 modification for the Fw 190A-8, F-3 and F-8; L, the R6 modifications of two Wfr. Gr.21 mortars on the Fw 190A-4, A-5, A-6 and A-7

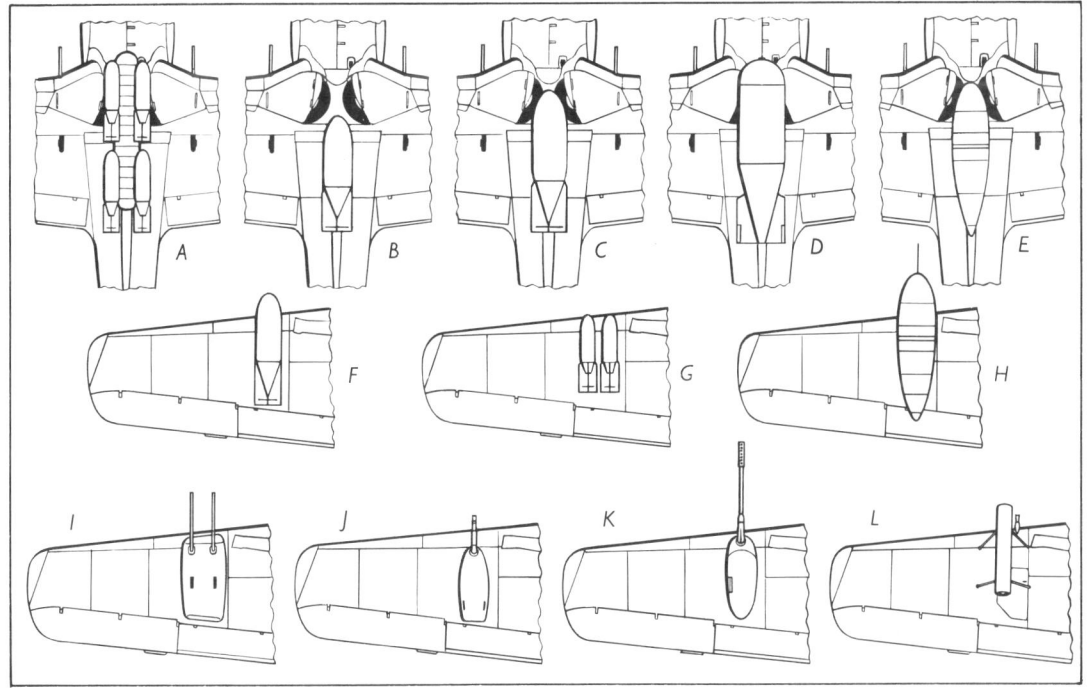

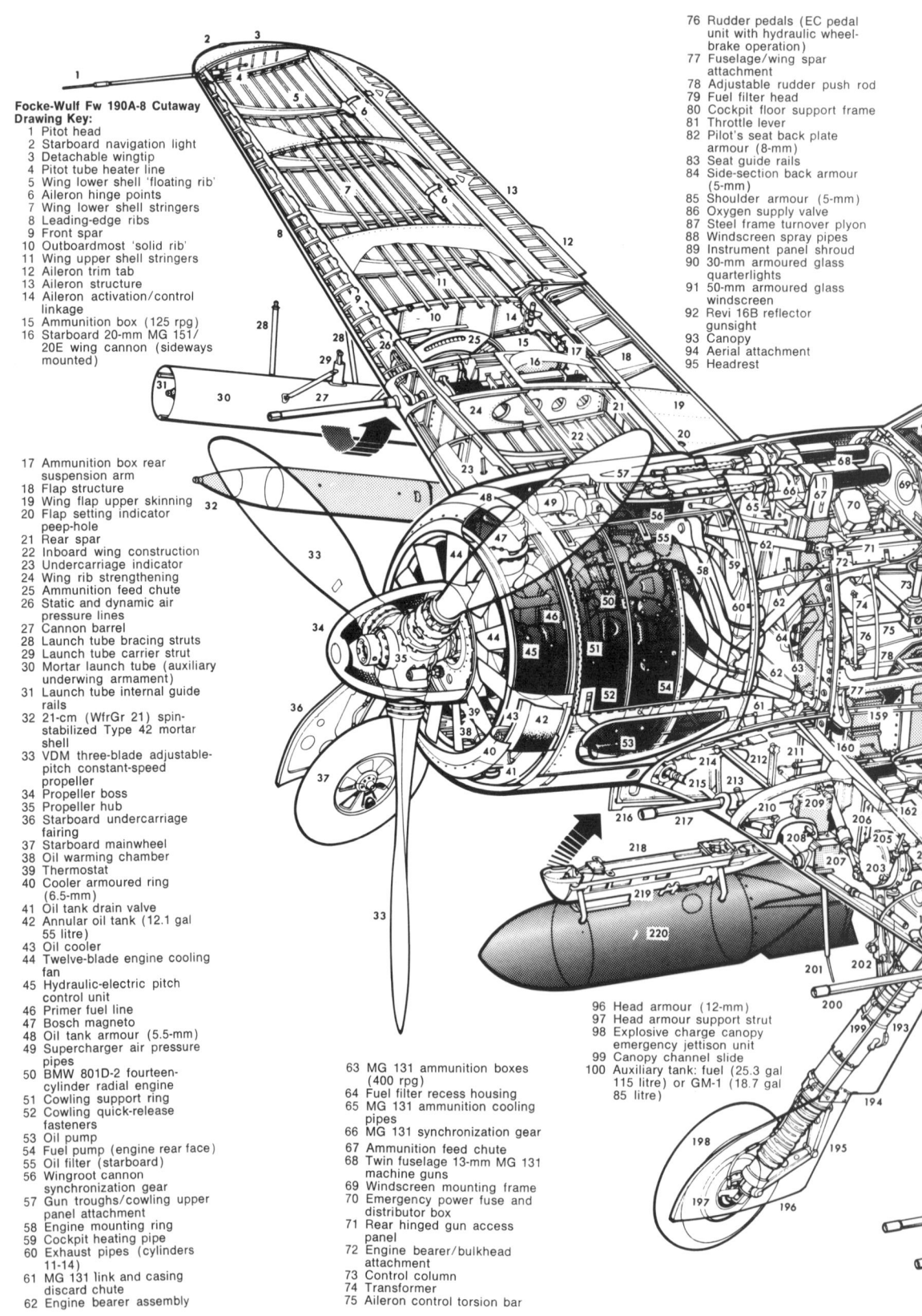

Focke-Wulf Fw 190A-8 Cutaway Drawing Key:
1. Pitot head
2. Starboard navigation light
3. Detachable wingtip
4. Pitot tube heater line
5. Wing lower shell 'floating rib'
6. Aileron hinge points
7. Wing lower shell stringers
8. Leading-edge ribs
9. Front spar
10. Outboardmost 'solid rib'
11. Wing upper shell stringers
12. Aileron trim tab
13. Aileron structure
14. Aileron activation/control linkage
15. Ammunition box (125 rpg)
16. Starboard 20-mm MG 151/20E wing cannon (sideways mounted)
17. Ammunition box rear suspension arm
18. Flap structure
19. Wing flap upper skinning
20. Flap setting indicator peep-hole
21. Rear spar
22. Inboard wing construction
23. Undercarriage indicator
24. Wing rib strengthening
25. Ammunition feed chute
26. Static and dynamic air pressure lines
27. Cannon barrel
28. Launch tube bracing struts
29. Launch tube carrier strut
30. Mortar launch tube (auxiliary underwing armament)
31. Launch tube internal guide rails
32. 21-cm (WfrGr 21) spin-stabilized Type 42 mortar shell
33. VDM three-blade adjustable-pitch constant-speed propeller
34. Propeller boss
35. Propeller hub
36. Starboard undercarriage fairing
37. Starboard mainwheel
38. Oil warming chamber
39. Thermostat
40. Cooler armoured ring (6.5-mm)
41. Oil tank drain valve
42. Annular oil tank (12.1 gal 55 litre)
43. Oil cooler
44. Twelve-blade engine cooling fan
45. Hydraulic-electric pitch control unit
46. Primer fuel line
47. Bosch magneto
48. Oil tank armour (5.5-mm)
49. Supercharger air pressure pipes
50. BMW 801D-2 fourteen-cylinder radial engine
51. Cowling support ring
52. Cowling quick-release fasteners
53. Oil pump
54. Fuel pump (engine rear face)
55. Oil filter (starboard)
56. Wingroot cannon synchronization gear
57. Gun troughs/cowling upper panel attachment
58. Engine mounting ring
59. Cockpit heating pipe
60. Exhaust pipes (cylinders 11-14)
61. MG 131 link and casing discard chute
62. Engine bearer assembly
63. MG 131 ammunition boxes (400 rpg)
64. Fuel filter recess housing
65. MG 131 ammunition cooling pipes
66. MG 131 synchronization gear
67. Ammunition feed chute
68. Twin fuselage 13-mm MG 131 machine guns
69. Windscreen mounting frame
70. Emergency power fuse and distributor box
71. Rear hinged gun access panel
72. Engine bearer/bulkhead attachment
73. Control column
74. Transformer
75. Aileron control torsion bar
76. Rudder pedals (EC pedal unit with hydraulic wheel-brake operation)
77. Fuselage/wing spar attachment
78. Adjustable rudder push rod
79. Fuel filter head
80. Cockpit floor support frame
81. Throttle lever
82. Pilot's seat back plate armour (8-mm)
83. Seat guide rails
84. Side-section back armour (5-mm)
85. Shoulder armour (5-mm)
86. Oxygen supply valve
87. Steel frame turnover plyon
88. Windscreen spray pipes
89. Instrument panel shroud
90. 30-mm armoured glass quarterlights
91. 50-mm armoured glass windscreen
92. Revi 16B reflector gunsight
93. Canopy
94. Aerial attachment
95. Headrest
96. Head armour (12-mm)
97. Head armour support strut
98. Explosive charge canopy emergency jettison unit
99. Canopy channel slide
100. Auxiliary tank: fuel (25.3 gal 115 litre) or GM-1 (18.7 gal 85 litre)

40

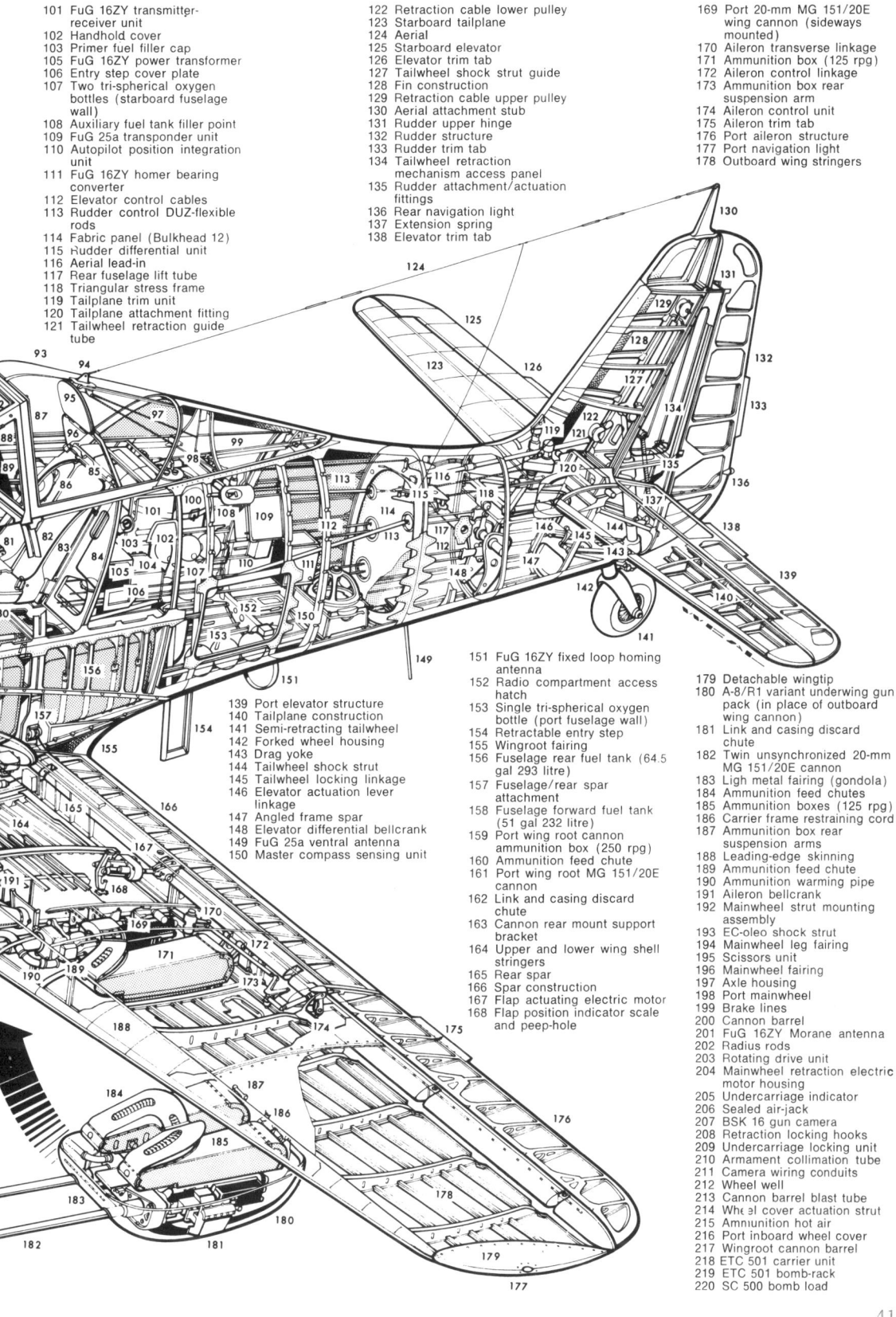

101 FuG 16ZY transmitter-receiver unit
102 Handhold cover
103 Primer fuel filler cap
105 FuG 16ZY power transformer
106 Entry step cover plate
107 Two tri-spherical oxygen bottles (starboard fuselage wall)
108 Auxiliary fuel tank filler point
109 FuG 25a transponder unit
110 Autopilot position integration unit
111 FuG 16ZY homer bearing converter
112 Elevator control cables
113 Rudder control DUZ-flexible rods
114 Fabric panel (Bulkhead 12)
115 Rudder differential unit
116 Aerial lead-in
117 Rear fuselage lift tube
118 Triangular stress frame
119 Tailplane trim unit
120 Tailplane attachment fitting
121 Tailwheel retraction guide tube
122 Retraction cable lower pulley
123 Starboard tailplane
124 Aerial
125 Starboard elevator
126 Elevator trim tab
127 Tailwheel shock strut guide
128 Fin construction
129 Retraction cable upper pulley
130 Aerial attachment stub
131 Rudder upper hinge
132 Rudder structure
133 Rudder trim tab
134 Tailwheel retraction mechanism access panel
135 Rudder attachment/actuation fittings
136 Rear navigation light
137 Extension spring
138 Elevator trim tab
139 Port elevator structure
140 Tailplane construction
141 Semi-retracting tailwheel
142 Forked wheel housing
143 Drag yoke
144 Tailwheel shock strut
145 Tailwheel locking linkage
146 Elevator actuation lever linkage
147 Angled frame spar
148 Elevator differential bellcrank
149 FuG 25a ventral antenna
150 Master compass sensing unit
151 FuG 16ZY fixed loop homing antenna
152 Radio compartment access hatch
153 Single tri-spherical oxygen bottle (port fuselage wall)
154 Retractable entry step
155 Wingroot fairing
156 Fuselage rear fuel tank (64.5 gal 293 litre)
157 Fuselage/rear spar attachment
158 Fuselage forward fuel tank (51 gal 232 litre)
159 Port wing root cannon ammunition box (250 rpg)
160 Ammunition feed chute
161 Port wing root MG 151/20E cannon
162 Link and casing discard chute
163 Cannon rear mount support bracket
164 Upper and lower wing shell stringers
165 Rear spar
166 Spar construction
167 Flap actuating electric motor
168 Flap position indicator scale and peep-hole
169 Port 20-mm MG 151/20E wing cannon (sideways mounted)
170 Aileron transverse linkage
171 Ammunition box (125 rpg)
172 Aileron control linkage
173 Ammunition box rear suspension arm
174 Aileron control unit
175 Aileron trim tab
176 Port aileron structure
177 Port navigation light
178 Outboard wing stringers
179 Detachable wingtip
180 A-8/R1 variant underwing gun pack (in place of outboard wing cannon)
181 Link and casing discard chute
182 Twin unsynchronized 20-mm MG 151/20E cannon
183 Ligh metal fairing (gondola)
184 Ammunition feed chutes
185 Ammunition boxes (125 rpg)
186 Carrier frame restraining cord
187 Ammunition box rear suspension arms
188 Leading-edge skinning
189 Ammunition feed chute
190 Ammunition warming pipe
191 Aileron bellcrank
192 Mainwheel strut mounting assembly
193 EC-oleo shock strut
194 Mainwheel leg fairing
195 Scissors unit
196 Mainwheel fairing
197 Axle housing
198 Port mainwheel
199 Brake lines
200 Cannon barrel
201 FuG 16ZY Morane antenna
202 Radius rods
203 Rotating drive unit
204 Mainwheel retraction electric motor housing
205 Undercarriage indicator
206 Sealed air-jack
207 BSK 16 gun camera
208 Retraction locking hooks
209 Undercarriage locking unit
210 Armament collimation tube
211 Camera wiring conduits
212 Wheel well
213 Cannon barrel blast tube
214 Wheel cover actuation strut
215 Ammunition hot air
216 Port inboard wheel cover
217 Wingroot cannon barrel
218 ETC 501 carrier unit
219 ETC 501 bomb-rack
220 SC 500 bomb load

41

Above and below, contrasting views of a *Jabo* version of the Fw 190 and a *Zerstörer* variant. The upper photo shows the wing and fuselage bomb-racks; no outer wing guns are fitted and the pitot head is near the wing tip; a tropical filter is fitted. The aircraft below is the second Fw 190A–5/U9, *Werk-Nr* 816, with MG 131 machine guns in the upper fuselage and MG 151s in the wing roots and the outer wing bays

and air-defence capability. Associated with these developments was a redesigned wing structure, made necessary by the steadily increasing gross weight of the aircraft and the greater stresses imposed by the heavier wing armament now being considered. This new wing, with a slight increase of span, became a production feature effective with the Fw 190A–6 sub-series but was first flown on the two A–5/U9s and two A–5/U10s. The former (*Werk-Nummern* 812 and 816) were the first to incorporate 13-mm MG 131 machine guns in the fuselage, in place of the 7.9-mm MG 17s in all previous examples; two 20-mm MG 151s were retained in the wing root installation, and on the second aircraft, which was later redesignated the Fw 190 V35, MG 151s were also substituted for the outer-wing MG FF guns. The A–5/U10s (*Werk-Nummern* 861 and 862) were in effect the prototypes for the A–6, having the new wing with MG 151s in inner

Above, a close-up of the tray containing two MG 151 cannon, carried beneath each wing of the Fw 190A–5/U12 variant

Above right and below, two views of the Fw 190A–5/U12, *Werk-Nr* 813, one of the two prototypes tested in this configuration in the summer of 1943 in the search for a more effective ground-attack version

Above and below, three views of the Fw 190A–5/U14, *Werk-Nr* 871, a variant of the Focke-Wulf fighter adapted to carry an LTF 5b torpedo on the ETC 502 fuselage rack. Also obvious in these views are the enlarged fin, with broader chord at the top, and lengthened tail wheel leg

and outboard positions, and MG 17 fuselage guns.

To enhance the ground-attack effectiveness of the Fw 190, the installation of a 30-mm MK 103 cannon beneath each wing was next investigated, in the single Fw 190A–5/U11 (*Werk-Nr* 1302); these cannon took the place of the outer MG FF guns, the usual MG 17s and MG 151s being retained, plus the fuselage centreline racks. A variation on the same theme was to fit, beneath each wing, a tray containing two 20-mm MG 151 guns side by side, this installation being known as the WB 151/20. *Werk-Nummern* 813 and 814 were converted to test this armament in the summer of 1943 as the Fw 190A–5/U12s.

The Fw 190A–5/U13 was a further exploitation of the *Jabo-Rei* theme, combining features of the U2 for night flying and the U8 armament configuration, with new Focke-Wulf Mtt underwing racks able to carry either one SC 250 (550-lb) bomb or one 300 litre (66 Imp gal) fuel tank. Three examples are believed to have been converted, including *Werk-Nr* 817 and possibly 825 and 1083.

An entirely different rôle had been envisaged for the Fw 190 during 1943, which accounted for the A–5/U14 and U15 designations. These were used to evaluate the practicability of using the fighter to carry and launch anti-shipping torpedoes. The ETC 502 fuselage rack was adapted to carry the LTF 5b torpedo, the length of which required that the tailwheel be lengthened to give adequate ground clearance and the fin had to be enlarged to provide satisfactory control. The fuselage and outer wing guns were

Above right and below, two views of the Fw 190A–5/U11, a *Schlachtflugzeug* fitted with long-barrel MK 103 cannon of 30-mm calibre in fairings beneath the wing to serve as a prototype for the *Rüstsatz* R3 modification on the later production versions

45

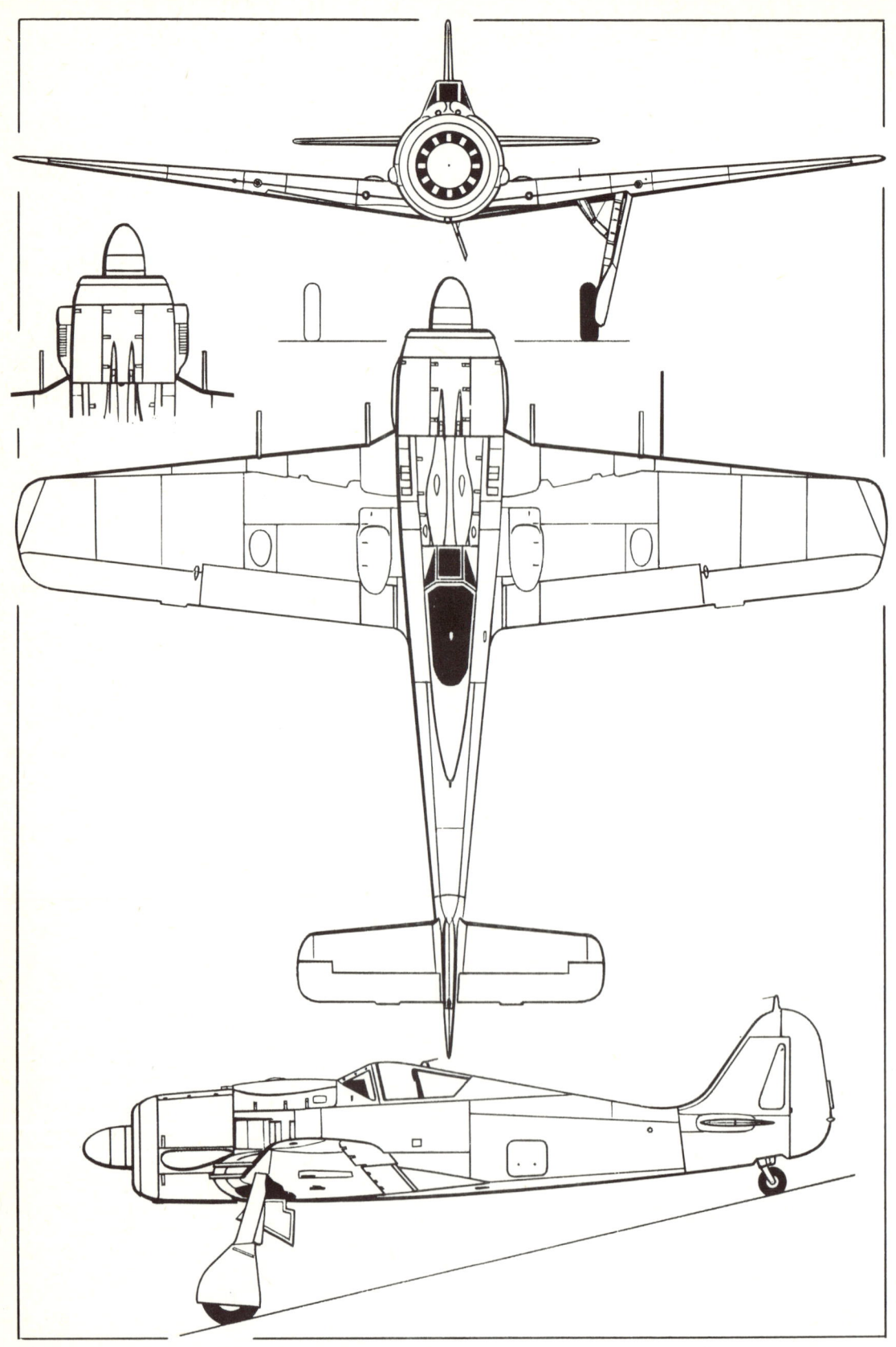

deleted, two MG 151s being retained in the roots, and provision was made for underwing fuel tanks to be carried. Two trials aircraft (*Werk-Nummern* 871 and 872) were modified to this A–5/U14 configuration, the gross weight rising to 11,464 lb (5 000 kg). The single A–5/U15, believed to be *Werk-Nr* 1282, was similar, but it was adapted to carry the Blohm und Voss LT950 guided torpedo with Askania ALSK 121 guidance equipment, and had underwing fuel tanks.

Continuing the search for improved armament in the anti-bomber rôle, which by mid-1943 was becoming a major commitment for the Fw 190 force, Focke-Wulf installed a 30-mm MK 108 cannon beneath each wing of the single Fw 190A–5/U16 (*Werk-Nr* 1346), replacing the outer wing MG 151 guns. The single A–5/U17 was another *Schlachtflugzeug* development aircraft, leading to production of the Fw 190F–3 in due course, armed with two MG 17s in the fuselage, two MG 151s in the wing roots and an ETC 50 rack beneath each wing. The use of an MW 50 methanol-water power-boost system was projected for a trial in the A–5/U18, but this appears not to have materialised. Ten examples of the Fw 190A–5k were built to investigate the use of duralumin in the structure.

As already noted, the Fw 190A–6 differed from the A–5 primarily in having a revised wing structure, the basic armament comprising two MG 17s in the fuselage and two MG 151s each in the wing roots and outer wing installation. Additional armour protection was provided, and the A–6 was intended primarily for service on the Eastern front as a *Zerstörer* (destroyer). It followed the A–5 on to the production lines in the spring of 1943, and was the subject of several *Rüstsatz* schemes that varied the armament to meet specific operational needs. Thus, the A–6/R1 had the outer wing guns removed and carried instead two WB 151/20 packs, one beneath each wing, increasing the total armament to six MG 151s plus the two MG 17s in the fuselage. In the A–6/R2, a 30-mm MK 108 weapon was fitted in each wing in place of the outer MG 151, this installation having been developed on the Fw 190 V51 (*Werk-Nr* 765). The designation A–6/R4 was used for

Left, a three-view drawing of the Fw 190A–8, one of the principal versions of the Focke-Wulf fighter and one that was built in larger numbers than any other sub-type. Two sub-variants of the A–8 are depicted below: the Fw 190A–8/U1 two-seat conversion trainer and the A–8/R3 with 30-mm MK 103 cannon in underwing fairings

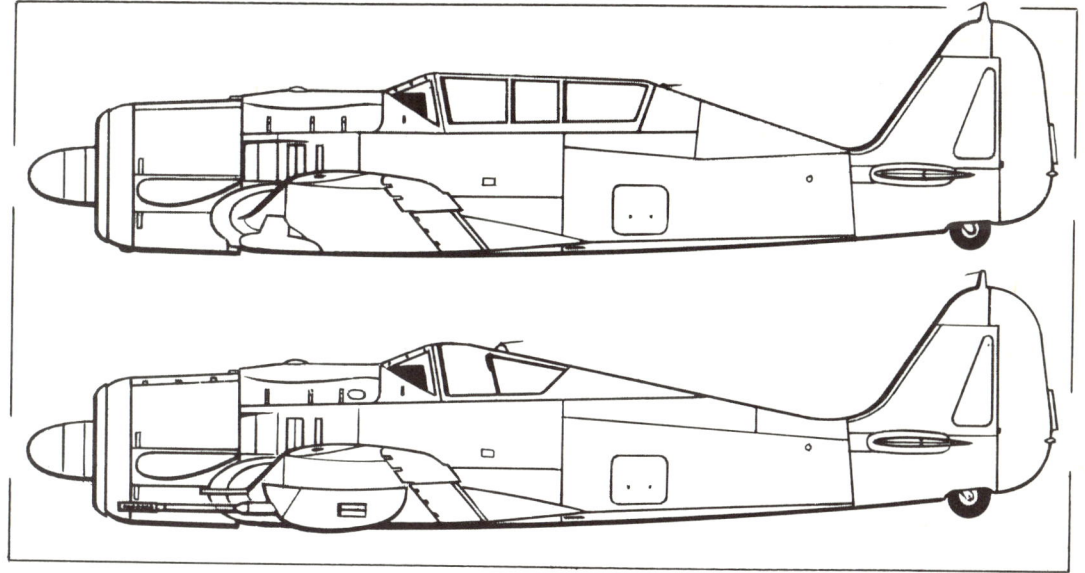

Another view of the Focke-Wulf Fw 190A-5/U9, *Zerstörer* (also shown on page 42), the prototype used to test the new wing that became effective on the Fw 190A-6 production version

a version having a GM 1 power-boat system, tested in the Fw 190 V45 (*Werk-Nr* 7347) and V47 (*Werk-Nr* 530115) prototypes but believed not to have been deployed operationally until the advent of the Fw 190A-8/R4. Some R6 kits, comprising the Wfr Gr 21 morters, were fitted to A-6 airframes as well as to A-4s and A-5s, as already mentioned.

Production records indicate that 569 of the A-6 sub-variants were built, bringing the total by late 1943 to 3,223 (in addition to more than 1,300 of the related Fw 190F and G series, described in later pages). During 1943 alone the Luftwaffe took delivery of 3,208 Fw 190s from Focke-Wulf, AGO, Arado and Fieseler, of which 325 were accepted in July. These figures represented about one-third of all German single-seat fighter production at that time. Three more sub-series of the Fw 190A were still to appear, of which one, the A-7, was built in only small numbers; another, the A-8, was to be built in larger numbers than any other sub-type, and the last, the A-9, was to remain obscure.

Although only 80 Fw 190A-7s were built, examples appear to have been completed on three of the four production lines, Arado being the exception. New features of this series were simplified electrics and a Revi 16b gunsight in place of the previous C12d.

Basic armament comprised two MG 131s in the fuselage and four MG 151s in the wings, but about half the total built were modified to Fw 190A-7/R2 with two MK 108s replacing the two outer MG 151s and thus being equivalent to the A-6/R2. Some Fw 190A-7/R6 versions also appeared in service, with the underwing mortars.

Whereas the Fw 190A-6 and A-7 emerged as variants of the A-5 largely as a result of expedient, the A-8 was evolved in the Focke-Wulf design office as the next major improvement of the A series, again taking the A-5 as the starting point. Powered by a BMW 801D-2 engine, it had a basic armament of two MG 131s in the fuselage and four MG 151s in the wings. Provision was made in the fuselage aft of the cockpit for an auxiliary fuel tank of 115 litre (25 Imp gal) capacity (which could be replaced by a GM 1 nitrous oxide tank), and to keep the CG within required limits, the FuG 16Z-Y radio (replacing the earlier FuG 16Z-E set, from which it differed in having an additional homing facility) was moved forward, as also was the ETC 501 fuselage centreline bomb-rack (by 7·9 in/20 cm). Use of the GM 1 system in the Fw 190A-8/R4 version permitted higher powers to be achieved at altitudes above 26,250 ft (8 000 m), with a consequent speed increase of about 36 mph (58 km/h). Some

Fw 190A–8 Specification

Power Plant: One BMW 801D–2 14-cylinder radial air-cooled engine rated at 1,700 hp for take-off and 1,440 hp at 18,700 ft (5 700 m). Fuel capacity, 115.5 Imp gal (524 litres) in two fuselage tanks, plus 25.3 Imp gal (115 litres) in optional rear fuselage tank plus provision for 66.2 Imp gal (300 litre) drop tank.

Performance (Clean): Max speed, 355 mph (571 km/h) at sea level, 402 mph (647 km/h) at 18,045 ft (5 500 m); max speed with GM 1 nitrous oxygen boost, 408 mph (656 km/h) at 20,670 ft (6 300 m); normal cruising speed, 298 mph (480 km/h) at 6,560 ft (2 000 m); initial rate of climb, 3,450 ft/min (17,5 m/sec); time to climb to 19,685 ft (6 000 m), 9.1 min; to 26,250 ft (8 000 m), 14.4 min; to 32,800 ft (10 000 m), 19.3 min; service ceiling, 33,800 ft (10 300 m) and with GM 1 boost, 37,400 ft (11 400 m); max range, 644 mls (1 035 km) at 22,970 ft (7 000 m); range with one drop tank, 915 mls (1 470 km) at 301 mph (485 km/h) at 16,400 ft (5 000 m).

Weights: Empty equipped (clean), 7,652 lb (3 470 kg); empty equipped (fighter-bomber), 7,740 lb (3 510 kg); normal loaded, 9,660 lb (4 380 kg); max take-off (fighter-bomber), 10,724 lb (4 865 kg).

Dimensions: Span, 34 ft 5½ in (10,506 m); length, 29 ft 4¼ in (8,95 m); height (over airscrew), 12 ft 11½ in (3,95 m); wing area, 196.98 sq ft (18,3 m²); undercarriage track, 11 ft 6 in (3,50 m).

Armament: Two 13-mm MG 131 machine guns with 475 rpg in fuselage; two 20-mm MG 151–20E cannon with 250 rpg in wing roots and two 20-mm MG 151–20E cannon with 140 rpg in outer wing panels.

A close-up of a *Jabo* version of the Fw 190, showing two SC 50 bombs beneath each wing and four similar bombs in tandem pairs beneath the fuselage.

One of the few Fw 190A-8/U1 two-seat trainers, Experience showed that few pilots needed instruction on such a two-seat version and plans for a production version were abandoned

Fw 190A-8s also had an 'emergency power unit' that allowed the supercharger boost regulator to be overridden for up to 10 minutes at a time, increasing the speed by about 5 mph (25 km/h).

Production of the Fw 190A-8 totalled 1,334 aircraft and was shared between Focke-Wulf, AGO, Fieseler and Dornier, the last-mentioned company being brought into the programme in 1944 and building the Focke-Wulf fighter at Wismar. Also during 1944 arrangements were made for the Fw 190 to be built in an underground plant operated in France by the SNCA du Centre at Cravant, near Auxerre. The first of the fighters to emerge from this factory, where the will to produce weapons for the Luftwaffe was not noteworthy, was an Fw 190A-5, flown for the first time on 16 March 1945. In the next 12 months sixty-four aircraft of the A-5 and A-8 series were built in France, some of these entering service with the GC III/5 *Normandie-Niemen* squadron of the *Armée de l'Air* for a brief period with the designation NC 900.

The first two NC 900s reached GC III/5 in October 1945 and with subsequent deliveries they were used to equip the 4eme *Escadrille*.

Rüstsätze resulted in Fw 190A-8/R1 and R2 variants going into service, with the same respective armament as the equivalent Fw 190A-6 variants. The A-8/R3 differed from the R2 in having 30-mm MK 103 cannon instead of MK 108s in the outer wing bays, this weapon having a longer barrel, lower rate of fire and higher muzzle velocity.

The intensity of the bombing attacks over Germany by 1944 brought about a new concept in air fighting involving the Fw 190, the A-8/R7 version of which was specially strengthened to make possible ramming attacks against the B-17s and B-24s of the US Eighth Air Force. Modifications were primarily concerned with the armour protection around the cockpit, and these aircraft, known as *Rammjäger*, were used by the specially formed *Sturmstaffeln* (attack squadrons) under Major Dahl. The Fw 190A-8/R7 carried standard armament, whereas the A–8/R8, also a *Rammjäger*, was to have the R2 standard of armament; the latter, however, did not reach operational status. Some work was also done by Focke-Wulf on a suicide version (*Selbstvernichtung-* or *SV-Flugzeug*) intended to dive into bomber formations, one Fw 190A–5/U8 and one Fw 190A–8 being assigned for tests in this rôle. Schemes were also drawn up, but apparently not implemented, to produce Fw 190A–8/R11 and R12 all-weather fighters, respectively with standard and R2 armament, the new features being the use of a turbosupercharged **BMW 801TU** engine, PKS 12 directional control system, heated front and port windscreen panels and FuG 125 radio range approach system. The Fw 190 V72 (*Werk-Nr 170727*), tested in August 1944, incorporated these features. The BMW 801TU engine had been developed as an intermediate step between the BMW 801D–2 and the later, more powerful, BMW 801TS or TH; it had certain drive components of the latter model and heavier armoured nose rings.

The continuing search for more performance and improved fighting capability was reflected in a number of experiments conducted with Fw 190A–8 airframes during 1944, one such being the Fw 190 V45 proto-

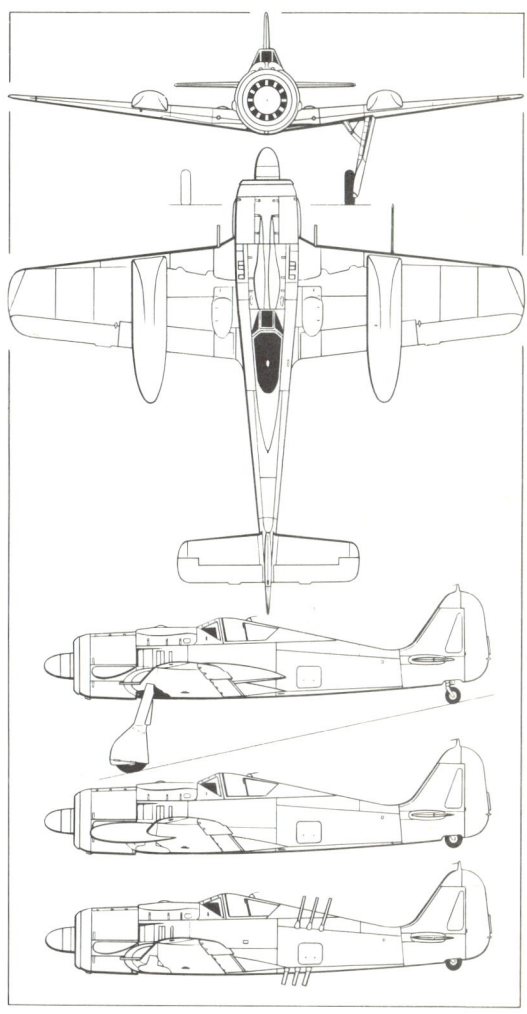

Above right, a three-view drawing of the Fw 190A–8 *Doppelreiter* I (Double-rider) with auxiliary fuel tanks on the upper wing surfaces and, central side view, the *Doppelreiter* II with tanks on the leading edge. The bottom side view shows an Fw 190A–8 fitted with an SG 116 *Zellendusche* salvo-weapon, designed to fire 30-mm shells vertically into bomber formations and triggered by photo-electric cells. Below, Fw 190As built in France by SNCA du Centre and serving, with the designation NC 900, with the GC III/5 *Normandie-Niemen* squadron

Above and below, two views of the Fw 190A–8/V26 modified as a carrier for a Blohm und Voss Bv 246 *Hagelkorn* (Hailstone) gliderbomb. Vertical bars attached beneath the wings of the Fw 190 exerted a pressure on the wings of the Bv 246 to project the latter away from its carrier aircraft when released

type with a long-span wing (designed by SNCA du Sud-Ouest at Chatillon-sur-Seine and having a span of 40 ft $4\frac{1}{4}$ in/12,30 m) to improve high altitude performance. Another A–8 was used for an extensive series of trials with the so-called *Doppelreiter* (Double-rider) auxiliary fuel tanks on the upper wing surfaces and extending aft in a configuration similar to the drag-reducing fairings used after the war on the Convair 990 transport and Handley Page Victor bomber. These tanks had been evolved by the *Forschungsgruppe Graf Zeppelin* (Zeppelin Research Establishment) and had a capacity of 250 litres (55 Imp gal) each; although their drag characteristics were much better than those of underwing tanks, they were found to impair the handling of the Fw 190, making it so unstable that their adoption for service use appeared unwise.

Among the most interesting armament concepts tested on the Fw 190A was the vertically mounted *Zellendusche* salvo-weapon. In the SG 116 version, tested on an Fw 190A–8, this comprised three 30 mm MK 103 cannon barrels mounted near vertically in the fuselage just behind the cockpit; each contained one shell and a 'magic eye' photo-electric cell (*Foto-Zellenfühler*) which would detect the shadow of a target overhead and fire the weapon. The barrels were so mounted as to give a spread of 1½ deg between the foremost and rearmost, and a timing device sequenced the operation to give 3/100ths of a second separation between each shell. The similar SG 117 installation used double the number of MK 108 cannon. Another Fw 190A–8 was tested with an aft-firing 21-cm rocket tube carried horizontally beneath the fuselage; and yet another, redesignated as the Fw 190 V69, carried an X4 *Ruhrstahl* (DVL 344) wire-guided rocket missile beneath each wing. The so-called *Rohrblock* 108 armament installation, comprising a pack of seven MK 108 barrels carried beneath each wing, was to have been tested on an A–8 redesignated as the Fw 190 V74. The Blohm und Voss BV 246 *Hagelkorn* (Hailstone) glider bomb was tested from an Fw 190A–8/U2b.

If some of these experiments and concepts seemed far-fetched, then another variation based on the Fw 190A–8 was both mundane and practical: this was the A–8/U1 two-seat trainer. The need for such a conversion was primarily in relation to the increasing availability of the *Schlachflugzeug* versions of the Fw 190 to re-equip *Geschwader* operating the Ju 87s, and at one period during 1943 one *Gruppe* was converting every three weeks. There were fears in some Luftwaffe quarters that the Ju 87 pilots might find difficulty converting to the higher-performing Fw 190 and the training variant was therefore evolved, with a second cockpit, including only rudimentary dual controls and instrumentation for the instructor, located immediately behind the first. A long

Operating with I/SG 4 in Italy in 1943/44, this Fw 190F-7 *Jabo* has had some of its national insignia painted over, including the tail swastika and upper wing crosses, to achieve a better camouflage

continuous canopy was fitted, with a sideways opening section for the pupil to gain access to the front cockpit, and a rearwards-sliding cover over the rear cockpit; all armament was deleted. The first of three Fw 190A–8/U1s flew on 23 January 1944 and thereafter some Fw 190A–5s and A–8s were converted in the field as *Schulflugzeug*, taking the revised designations of Fw 190S–5 and Fw 190S–8; but the need for such trainers in fact proved less than had been anticipated.

In August 1944 a new prototype of the Fw 190 was transferred from the Focke-Wulf test establishment to Rechlin for Lufwaffe trials. This was the Fw 190 V34, apparently an A–8 airframe fitted with a 2,000 hp BMW 801F–1 powerplant and heavily armoured wing leading edge, for use as a *Rammjäger*. It was intended as the prototype of a new production series, the Fw 190A–9, which was basically the Fw 190A–8 with the turbosupercharged BMW 801TS or TU engine; only prototypes of this variant were completed, however, comprising the Fw 190 V35, V36, V72, V73 and V74. Variants were designated with the *Rüstsätze* R1, R2, R3, R8, R11 and R12. The Fw 190A–10 was to have been a *Jabo-Rei* variant with the BMW 801F–1 engine and a longer-span wing, but was not produced.

Down to Earth

Chronologically, the next series of the Focke-Wulf fighter after the Fw 190A was the Fw 190B, the most significant innovation in which was the use of a pressure cabin. This, and the related Fw 190C, are described later (p 65), however, since these – and the long-nosed Fw 190D – represented a greater technical step forward than was the case for the Fw 190F and G described in this chapter. The two last-mentioned were, in fact, Fw 190A airframes with no greater changes than had already been covered by the application of sub-series numbers, and the use of new series letters seems to have

The Fw 190F series was developed to give the 'Butcher Bird' the best possible ground attack characteristics, with wing and fuselage bomb racks taking precedence over built-in wing guns, and additional armour protection for the engine, oil tanks and pilot. Illustrated below is an Fw 190F–8

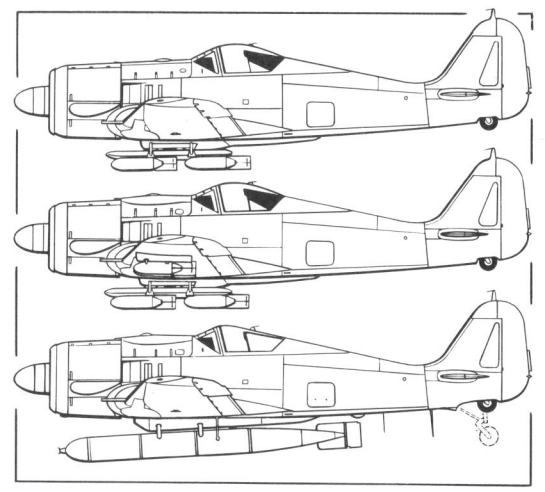

Three variants of the Fw 190F series shown in these side views were the F-2, based on the A-5 airframe and having a new one-piece blown canopy; the F-8 based on the A-8 airframe and the F-8/U14 adapted to carry a torpedo under the fuselage.

been more a matter of convenience than of logic. They were part of a trio of production designs intended to take advantage of *Umbau* modifications already evolved to suit the Fw 190 for reconnaissance, ground attack and long-range fighter-bomber duties. The first of this trio was intended to be the Fw 190E, referred to as a *Gefechtsaufklarer* and being an improvement of the Fw 190A-5/U4 with greater flexibility of camera installation; it was not produced.

The Fw 190F was a production version of the Fw 190A-5/U3 *Schlachtflugzeug*, although the first twenty-five to thirty examples were in fact based on the Fw 190A-4 and designated Fw 190F-1. Armament comprised two MG 17s in the fuselage and two MG 151s in the wing root; no outer wing guns were fitted, but one SC 250 (550-lb) bomb could be carried under each wing and the fuselage fitting carried either one ETC 501 rack for a single SC 500 (1,102-lb) bomb, or an ER 4 adapter which allowed four SC 50 (110-lb) bombs to be carried. Additional protective armour was provided for the engine, oil tank and pilot, and the undercarriage was strengthened for higher take-off weights. The Fw 190F-2 was similar, but based on the A-5 airframe with the new, slightly longer engine installation. A new one-piece blown cockpit canopy was also introduced, as the standard canopy on the Fw 190 had drawn some criticism from ground-attack pilots.

In mid-1943, Arado introduced the Fw 190F-3, based this time on the A-6 airframe with revised wing structure. This also had provision for a fuel tank on the fuselage rack, and in the Fw 190F-3/R1 version

Fw 190F-3 Specification

Power Plant: One BMW 801D-2 14-cylinder air-cooled radial engine rated at 1,700 hp for take-off and 1,440 hp at 18,700 ft (5 700 m).

Performance: Max speed (clean), 342 mph (550 km/h) at sea level, 394 mph (634 km/h) at 18,045 ft (5 500 m); max speed (with one 550-lb/250-kg bomb), 326 mph (525 km/h) at sea level, 368 mph (592 km/h) at 19,045 ft (5 500 m); initial rate of climb (clean), 2,110 ft/min (10,7 m/sec); max range (clean), 466 mls (750 km) at 296 mph (476 km/h) at 23,000 ft (7 000 m); range with one 550-lb/250-kg bomb), 330 mls (531 km) at 332 mph (534 km/h) at 18,045 ft (5 500 m).

Weights: Empty equipped, 7,328 lb (3 325 kg); normal loaded, 9,700 lb (4 400 kg); max take-off, 10,850 lb (4 920 kg).

Dimensions: Span, 34 ft $5\frac{1}{2}$ in (10,506 m); length, 29 ft $4\frac{1}{4}$ in (8,950 m); height (over airscrew), 12 ft $11\frac{1}{2}$ in (3,95 m); wing area, 196.98 sq ft (18,3 m²).

Armament: Two 7.9-mm MG 17 machine guns in fuselage and two 20-mm MG 151 cannon in wing roots. One fuselage centre-line bomb rack (ETC 250) for one 550-lb (250-kg) SC 250 bomb and (Fw 190F-3/R1) four underwing racks (ETC 50), each for one 110-lb (50-kg) SC 50 bomb or (Fw 190F-3/R3) two underwing 30-mm MK 103 cannon.

could carry four ETC 50 racks under the wings, to accommodate four 50-kg (110-lb) bombs. About 20 Fw 190F-3/R3s carried a 30-mm MK 103 cannon under each wing in place of the underwing bomb racks. The designations Fw 190F-4, F-5 and F-6 were applied to projected versions but were superseded by the F-8, F-9 and F-10, based on the Fw 190 A-8 airframe and built by Arado and Dornier during 1944. Of these, the Fw 190F-8 differed from the F-3 in having MG 131s in the fuselage and the ETC 50 racks under the wings as standard equipment. Production totalled 385, some being equipped for tropical use as the Fw 190F-8tp.

Provision for carrying underwing fuel tanks was made on the Fw 190F-8/U1, while the U2 and U3 versions were intended to carry the highly promising BT (*Bomben-Torpedo*) weapons that were being developed by the FGZ at Stuttgart. The BT was intended as an anti-shipping weapon and was more streamlined than the conventional LT torpedo; it had the advantage that its trajectory did not deviate when it hit the water and it was available in several sizes, including the 700-kg (1,543-lb) BT 700 and the 1,400-kg (3,088-lb) BT 1400, which were to be carried by the Fw 190F-8/U2 and U3

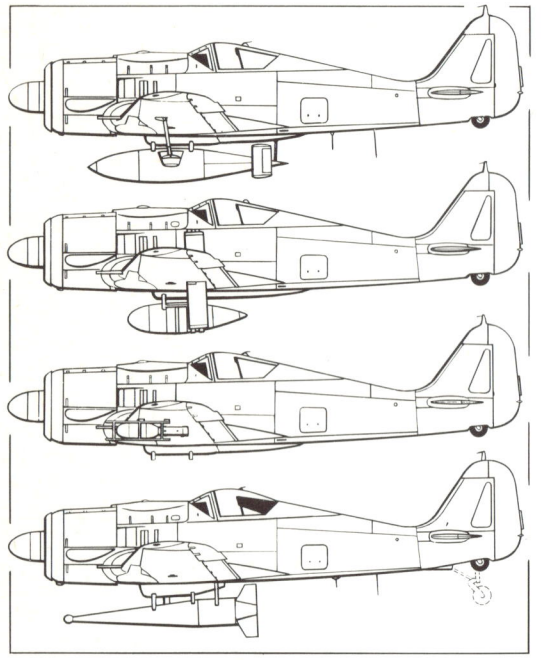

Above, experimental armament installations carried by Fw 190s included, top to bottom, the Bv 238 glider bomb on the A-8/U26; SG 113A *Forstersonde* anti-tank 77-mm recoilless guns on an F-8; 280-mm *Werfer-Granate* 28/32 anti-tank projectiles on an F-8 and a 1,543-lb (700-kg) BT 700 *Bomben-Torpedo* on an F-8/U2. Below, Fw 190F-8s carrying 550-lb (250-kg) bombs set out for a mission on the Eastern front in 1944

Above, a captured Focke-Wulf Fw 190F–8 in the Mediterranean theatre, with modified national markings. Below, a close-up of the SG 113A installation on an Fw 190F–8, comprising a pair of 77-mm recoilless guns in each wing, firing vertically downwards for anti-tank operation and triggered by the target's own electro-magnetic field

respectively, on an ETC 502 rack under the fuselage. An elongated tailwheel leg had to be fitted, together with a folding lower fin on the BT, to provide adequate ground clearance.

Trials were made with considerable success at the *Waffenprüfplatz* at Hexengrund in August 1944, and the suitability of the BT for use against land targets was also demonstrated. A special Luftwaffe unit was established to operate with this new weapon, using personnel drawn from I/NSG 5 and designated III/KG 200 in November 1944, and equipped with unmodified Fw 190F–8s for initial training. Plans were drawn up for the Blohm und Voss company to produce variants of the Focke-Wulf fighter equipped to carry the BT as the Fw 190F–8/R15 and R16, but the war ended before this plan could be put into effect and there is no record of the BT being used operationally from Fw 190s. Blohm und Voss was also under contract to produce the Fw 190F–8/R13, a

Above, below and bottom left, three views of the Fw 190G-3 *Werk-Nr* 636 carrying 300-1 (66-Imp gal) drop tanks on Focke-Wulf wing racks, and a 1,100-lb (500-kg) SC 500 bomb on the fuselage rack. The Fw 190G series was developed as the definitive *Jabo-Rei* version

night assault aircraft intended to succeed the Fw 190A-5/U2. It was to have had an advanced version of the BMW 801 with turbosupercharger, ETC 503 wing racks for drop tanks, FuG 16ZS and FuG 25 radio, and night-flying exhaust dampers. No production took place. Another abortive production programme concerned the Fw 190F-8/R14, intended to carry an LT torpedo under the fuselage, like the Fw

190A-5/U14, and to be built by Weserflug (previously the Rohrbach factory) in Berlin. Dornier produced two prototypes of the Fw 190F-8/R3 in November 1944, these having, like the F-3/R3 version, underwing MK 103 cannon in place of bomb racks.

Like the Fw 190A-8, the F-8 served as a test vehicle for several experimental armament installations, including a reversed application of the SG 116 *Zellendusche*. This comprised three MK 103 barrels mounted in the rear fuselage to fire downwards as an anti-tank weapon, but trials by *Versuchs-Jagdgruppe* 10 showed a marked lack of success. No greater success attended the trials with five 15-mm MG HF/15 barrels mounted beneath each wing, but another installation, comprising a trio of 88-mm *Wehrmacht* rocket missiles under each wing, did reach the stage of operational use on the Eastern front in October 1944, with the code name of *Panzershreck* (Tank Terror). The ballistics of the rocket weapon were not very satisfactory, however, and it was superseded by the *Panzerblitz* (Tank Lightning) or Pb 1 weapon in December 1944, this having about twice the effective range. Eight Pb 1s were carried by the Fw 190F-8 in jettisonable crates under the wings, but could only be fired at speeds below 305 mph (491 km/h), making the aircraft rather vulnerable to ground fire. This problem was alleviated in part when the Pb 2 was introduced, comprising a modified 55-mm R4M air-to-air missile with a hollow-charge warhead, up to seven being carried beneath each wing.

Pursuing the anti-tank possibilities of the aircraft, one Fw 190F-8 was fitted with 280-mm Wfr Gr 28/32 projectiles, carrying one or a pair beneath each wing, this weapon having been evolved from the Wfr Gr 21 used earlier against bomber formations but tested unsuccessfully on the Eastern front. Another F-8, redesignated the Fw 190 V75,

Views of a tropicalized Fw 190G-8/R5 showing alternative loads under the fuselage. Above right, a 1,100-lb (500-kg) SC 500 is being carried while below four 110-lb (50-kg) SC 50s are seen on an ER4 adapter. In both cases, four SC 50 bombs are carried on individual pylons under the wings

Above, a factory-fresh Fw 190G–2, *Werk-Nr* 410258 and below an Fw 190G–3 operating on the Eastern front in 1942, with an SD 500 bomb under the fuselage

was fitted with the SG 113A *Forstersonde,* comprising a pair of 77-mm recoilless guns firing vertically downwards from the wings and triggered by the electro-magnetic field created by the tank. Two more Fw 190F–8s were subsequently fitted with this weapon for additional trials, and others were used to test-fire the wire-guided X4 *Ruhrstahl* and X7 *Rotkäppchen.* Two Fw 190F–8s used for X4 launching trials in 1944 were redesignated as the V69 and V70 prototypes, and three others were used without special designations until the X4 programme ended in February 1945.

Final production version of the F series was the Fw 190F–9, built by both Arado and Dornier in the second half of 1944 and featuring a 2,000 hp turbosupercharged BMW 801TS engine, with a potential boosted power of 2,270 hp and able to maintain 1,715 hp at 40,000 ft (12 200 m). The Fw 190F–10 was to have been powered, like the A–10, with the BMW 801F engine, but no examples came off the line before the war ended. The Fw 190F–15 was built only as a prototype, redesignated the Fw 190 V66; with an airframe basically similar to that of the A–8, it had the BMW 801TS or TH engine, an ETC 504 multi-purpose fuselage bomb-rack and a modified undercarriage with larger wheels and hydraulic, in place of electric, actuation. The Fw 190F–16 was to be similar, with the addition of a TSA2D sight for missile aiming; a prototype was designated to be Fw 190 V67 and an airframe was earmarked for conversion, but is thought not to have been completed.

Evolved in parallel with the Fw 190F, the Fw 190G was regarded as a definitive *Jabo-Rei* version, consolidating at the production stage the features that had been introduced in the *Umbau* 8 and 13 modifications of the Fw 190A. Thus, powered by the BMW 801D–2 engine, the Fw 190G series had a fixed armament of only two 20-mm MG 151 guns in the wing roots, with an ETC 501 fuselage rack providing for one SC 500 (2,102-lb) or one SC 250 (550-lb) bomb to be carried, for four SC 50 (110-lb) bombs on an ER 4 converter together with the two 300 litre (66 Imp gal) underwing tanks. The Fw 190G–1, production of which totalled fifty, was based on the A–4 airframe, whereas the Fw 190G–2 was based on the slightly longer A–5 series. By October 1943 these variants had been superseded in production by the Fw 190G–3, which introduced a PKS 11 auto-pilot, *Kuti-Nase* balloon cable cutters in the wing leading edge, modified bomb racks and, on at least some examples, a direct fuel injection system, in which 96 octane petrol was injected direct into the port air intake and provided an increase in sea level power from 1,700 hp to 1,870 hp.

Above right, an Fw 190A–5/U8, *Werk-Nr* 1286, carrying a 2,200-lb (1,000-kg) SB 1000 bomb with lower fin removed to allow ground clearance, and (below) an Fw 190G–2 with a 1,100-lb (500-kg) SC 500 bomb and MttS (Messerschmitt-Stützen) to carry the 300-1 (66-Imp gal) drop tanks

Operating in Rumania early in 1943, this Fw 190G-3 benefits from a portable servicing shelter, facilitating the work of ground crews in very low temperatures and inclement weather generally

Thanks to this latter feature, the Fw 190G-3 had a maximum speed at sea level of 356 mph (573 km/h), an improvement of 16 mph (26 km/h) compared with the unboosted model. Some Fw 190G-3tp variants were produced for tropical use.

The final production version of the Fw 190G series was designated Fw 190G-8 to indicate that it had the same airframe as the A-8, production continuing from September 1943 to February 1944. These aircraft, like the Fw 190A-8, had the repositioned radio and fuselage rack, with provision behind the pilot for the auxiliary 115 litre (25 Imp gal) fuel tank, or, in the Fw 190G-8/R4, a GM-1 nitrous oxide booster tank. An adapter allowed the ETC 501 fuselage rack to carry a single 300 litre (66 Imp gal) fuel tank in place of the bomb, when an SC 250 (550-lb) bomb was carried on each wing rack, but later production batches lacked wing racks altogether and carried just the single bomb beneath the fuselage. In the Fw 190G-8/R5, of which 148 examples were produced, ETC 50 wing racks allowed four SC 50 (110-lb) bombs to be carried beneath each wing. Like the Fw 190A-8/R1, the Fw 190G-8/R1 carried two MG 151 cannon beneath each wing to improve its ground-attack effectiveness.

During their operational career some Fw 190Gs were adapted to carry heavier bomb loads for special missions. These included the SB 1000 and SC 1000 (2,205-lb) bombs, the fins of which had to be modified to give adequate ground clearance beneath the fuselage, and eventually the SC 1800 (3,968-lb) weapon, which was used by NSG 20 from its Fw 190Gs for pinpoint attacks on bridge targets during 1945. To carry the latter bomb, the Fw 190s had to be fitted with special tyres and much equipment was removed to keep the take-off weight within permissible limits, but even so the take-off run required by these aircraft was little short of 4,000 ft (1 220 m). Among experimental weapons carried by the Fw 190G for test purposes were the BV 246 *Hagelkorn* (Hailstone) glider bomb and the SB 800 (1,764-lb) RS *Kurt* rolling mine-bomb.

The Search for Altitude

By the beginning of 1942 enough information on the operational aspects of the Fw 190A had become available to Kurt Tank's team of engineers in Bremen to allow them to assess both its advantages and its weaknesses. It was already earning an excellent reputation as a fighting machine, with good manoeuvrability, excellent handling qualities and an overall performance that put it ahead of its principal opponents. Its versatility was beginning to be appreciated, with the development of reconnaissance and fighter-bomber versions to serve alongside the basic interceptor. But in one respect the Fw 190 still had a serious shortcoming, that being the rapid fall-off in performance at higher altitudes. Although the BMW 801 was nominally rated to produce its best power at 23,000 ft (7 010 m), experience soon showed that the Fw 190 was less than satisfactory as a fighting machine above altitudes of about 20,000 ft (6 096 m).

Several ways of improving the Fw 190's performance at altitude were open to investigation, including the use of a turbosupercharger on the BMW 801 or substitution of a completely new engine with better high-altitude characteristics, coupled with the introduction of a pressure cabin and increased wing area. All these alternatives received the attention of the design team and by the middle of 1942 three variants had been designated to cover all eventualities: the Fw 190B with a pressure cabin, larger wing and turbosupercharger; the Fw 190C with a Daimler-Benz DB 603 liquid-cooled engine;

The Fw 190 V13, one of the original Fw 190A–0 airframes, *Werk-Nr* 0036, was the first of the Focke-Wulf fighters to have a Daimler-Benz DB 603 engine installation, becoming in effect a prototype for the Fw 190C

and the Fw 190D with a Junkers Jumo 213A liquid-cooled engine. Since the DB 603 had been developed by Daimler-Benz without official approval, its use was discouraged by the *Technische Amt*, which regarded the Jumo as having a more certain future, but Tank himself favoured the former unit and was by this time in a strong enough position to obtain approval for development of the Fw 190C around it. During the latter months of 1942 plans were completed for production of both the Fw 190B and the Fw 190C, but these were destined never to see fruition, only the Fw 190D of this trio of proposals reaching production status (see p 70).

First step in the evolutionary process was to install a rudimentary pressure cabin in one

Top left, above and below, three photographs of the Fw 190V18/U1, the first of the so-called *Kanguruh* prototypes with exhaust-driven turbo-superchargers in the large ventral fairing. Other features were the enlarged fin and four-bladed propeller

of the Fw 190A–0 airframes (*Werk-Nr* 0035), which was then redesignated Fw 190 V12 as the prototype of the Fw 190B series. This aircraft was fitted with the GM 1 boost system, apparently being only the second example to have it, following an experimental installation in the Fw 190A–0/U12. The nitrous oxide for the GM 1 system was stored in liquid form, under pressure, and provided a source of additional oxygen to assist combustion when injected into the engine, thus helping to maintain power at high altitudes. The cabin was made pressure-tight by sealing the fireproof bulkhead behind the engine, sealing the floor and sides and fitting a reinforced double-walled sliding canopy with hot-air circulation between the layers. The useful life of this particular trials aircraft appears to have been somewhat circumscribed and it had been scrapped by November 1942.

The first four of the additional batch of Fw 190 development airframes (ordered to follow the original 40 Fw 190A–0s) were next assigned to the Fw 190B programme, these being designated Fw 190B–0 (*Werk-Nummern* 0046, 0047, 0048 and 0049). They were similar in general to the Fw 190 V12, having the GM 1 system and pressurised cabin, and the first of the quartet also had an increased-span wing, the area of which was 218.5 sq ft (20,3 m²) compared with the standard figure of 196.98 sq ft (18,3m²). This wing was at one time regarded as the future standard for both the Fw 190B and Fw 190C production aircraft but appears to have been flown, in fact, only on *Werk-Nr* 0046 and on one of the Fw 190D prototypes. All four Fw 190B–0 aircraft were tested by Focke-Wulf pilots at Langenhagen, the airfield near Hanover that is now the site of that city's international airport, but recurring difficulties with the GM 1 system and the pressure cabin slowed progress. Brief trials were made by Luftwaffe pilots at Rechlin in May 1943, by which time it had been realised that the weight of the new

A close-up of the DB 603A installation in the Fw 190 V16, the third prototype in the C-series and, like the other two, a converted A–0 airframe.

systems, associated with the original BMW 801 engine – for which the hoped-for turbo-supercharger had not become available – left virtually no margin for armament to be carried, and interest in the Fw 190B waned. Production of the Fw 190B–1 was abandoned, although up to six are believed to have been completed (*Werk-Nummern* 811 to 816), and the proposed Fw 190B–1/R1 lead fighter and Fw 190B–2 did not proceed.

To test the DB 603 installation meanwhile, seven of the original batch of Fw 190A–0s had been set aside, these being joined later by eight of the later batch of trials aircraft. The first installation of the new engine was made in *Werk-Nr* 0036, which then took the designation Fw 190 V13. Although it was a Vee in-line engine, the DB 603A–0 was installed behind an annular radiator which gave the aircraft a radial-engined appearance; the oil cooler intake was located beneath and behind the cowling, giving the aircraft a distinctive new profile, and the length increased by 2 ft 2½ in (66 cm) to 31 ft 0½ in (9,46 m). Apart from the engine, the Fw 190 V13 was largely unchanged from the Fw 190A–0, and no armament was carried; it crashed on 30 July 1942. The second and third prototypes, the Fw 190 V15 (*Werk-Nr* 0037) and Fw 190 V16 (*Werk-Nr* 0038) were to all intents and purposes similar and were

Above, another of the prototypes in the Fw 190C series, this Fw 190 V30 had a pressure cabin in addition to the turbosupercharger in the ventral fairing. Below left, a three-view drawing of the Fw 190C and, bottom side view, the initial form of the Fw 190 V13 and V16

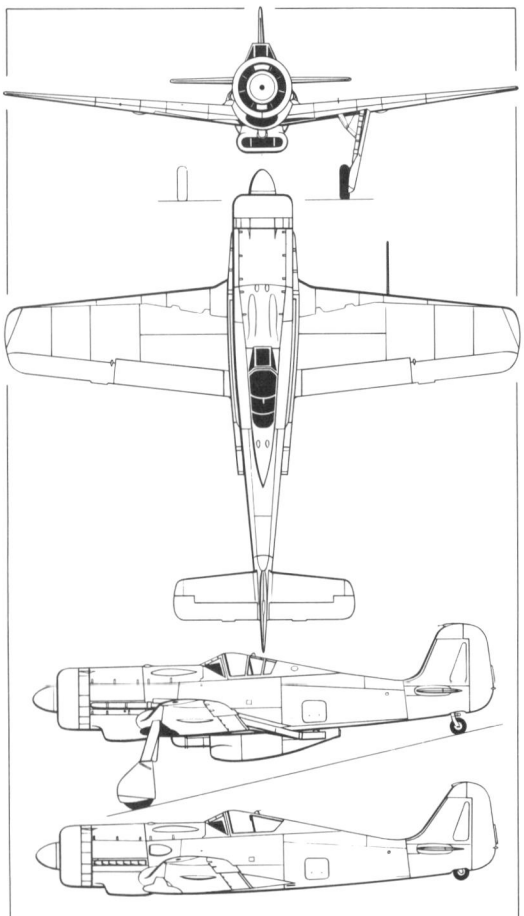

powered by the DB 603A, rated at 1,750 hp for take-off, 1,850 hp at 6,900 ft (2 100 m) and 1,625 hp at 18,700 ft (5 700 m).

Initially, none of these Fw 190C prototypes had turbosuperchargers fitted, but it was already clear that such a device would be essential if the operational altitudes of better than 45,000 ft (13 700 m) now being demanded by the *Technische Amt* were to be achieved. Meanwhile a supercharger installation was made in the Fw 190 V16, which was supplied to the Daimler-Benz *Erprobungsstelle* at Etchterdingen in August 1942 after Focke-Wulf trials had been completed at Langenhagen. To accommodate the D–B supercharger, the V16 was modified to have the oil cooler incorporated within the annular cowling, its intake being replaced by a larger one under the forward fuselage. With a DB 603E engine and increased-diameter supercharger, the Fw 190 V16 was able to reach an altitude of about 40,000 ft (12 200 m), and it also demonstrated, before the end of 1942, a speed of 450 mph (724 km/h) at 22,966 ft (6 800 m) and an initial rate of climb of 4,330 ft/min (22,0 m/sec). The DB 603E unit had a sea level output of 1,800 hp and a rating of 1,630 hp at 18,000 ft (5 500 m), but these two figures could be increased by use of the MW 50 methanol-water injection system, when they became 2,250 hp and 1,900 hp respectively.

Although these results were promising, they were not considered good enough, and attention was therefore focused upon the development of exhaust-driven turbosuperchargers which potentially could restore still more of the engine power in the rarified air of altitudes above 40,000 ft (12 200 m). Both the *Deutsche Versuchsanstalt für Luftfahrt* (DVL) and the Hirth-Motoren company were working in this field, although with difficulties because of the lack of steel alloys possessing adequate strength at high temperatures required for this application. Towards the end of 1942, however, the DVL was able to offer a prototype of its TK 11 (*Turbo-Kompressor* 11) for installation in an Fw 190C, the aircraft in question being the Fw 190 V18, another modified A–0 airframe (*Werk-Nr* 0040) that was redesignated Fw 190 V18/U1 when the TK 11 was fitted. With a DB 603A engine, it had a four-bladed VDM propeller, and the supercharger turbine was located in a

Above right, the Fw 190 V31 after it had crashed on 29 May 1943 on a flight from the test establishment at Rechlin. Below, *Dipl Ing* Kurt Tank, (centre, with unidentified members of his staff) in front of the Fw 190 V30 on the occasion of the first flight of the Ta 154 on 1 July 1943

The Fw 190 V21, one of the prototypes originally fitted with Jumo 213 engine and later converted to have the Daimler Benz DB 603. In both cases, an annular radiator was used for the in-line engine installation. It is shown here with the Jumo fitted

large ventral fairing, under the cockpit. Long ducts conducted the exhaust gases aft, over the wing roots, to feed into the turbine, from which compressed air was led forward through an intercooler to the supercharger itself.

The larger side area of the Fw 190 V18/U–1's fuselage called for more fin area to be provided, this being achieved by increasing the chord. The effect of the TK11 was to increase the power of the DB 603 from 1,560 hp at 24,200 ft (7 375 m) to 1,600 hp at 35,000 ft (10 700 m), but the unreliability of the pressure cabin imposed a considerable strain upon the pilots involved in the flight test programme at this time. More prototypes had meanwhile been put in hand, intended to lead to the production of the Fw 190C–1 (without pressure cabin) and Fw 190C–2 (with pressure cabin).

The first of these, Fw 190 V19 (*Werk-Nr* 0041), finally emerged with a major wing redesign, with the whole wing position moved forward by $4\frac{3}{4}$ in (12 cm), the leading edge at right-angles to the fuselage and the line of maximum thickness sweeping forward from root to tip. The enlarged fin developed for the Fw 190C series was fitted, and the rear fuselage was lengthened by 1 ft $7\frac{1}{2}$ in (50 cm). It was intended, when the design was finalised in April 1942, to use the DB 603A engine in the Fw 190 V19, but its construction was overtaken by the decision in 1943 to suspend the DB 603 variant in favour of the Jumo-engined Fw 190D. Consequently, this prototype, and six others with similar characteristics that had been laid down at the same time, were completed with Jumo 213 engines and became Fw 190D experimental models, as described on p 70.

Availability of the exhaust-driven supercharger, on the other hand, in the Fw 190 V18/U1 had led to the provision of five more prototypes from the second batch of development airframes, these being the Fw 190 V29 (*Werk-Nr* 0054), Fw 190 V30 (0055), Fw 190 V31 (0056), Fw 190 V32 (0057) and Fw 190 V33 (0058). All had the ventral fairing as developed for the Fw 190 V18/U1, and pressure cabins were fitted, although these contributed to the development problems encountered during flight tests. Basic engine for all these trials aircraft was the DB 603A driving either VDM or Schwarz four-bladed propellers; the Fw 190 V31 was destroyed on a test flight from Rechlin on 29 May 1943 and the other four were eventually fitted with Jumo 213 engines to serve as Fw 190D development aircraft, although *Werk-Nr* 0057, as the Fw 190 V32/U1 with the Hirth turbosupercharger deleted, was used for a time as a DB 603G test-bed by Daimler-Benz.

As production aircraft, the Fw 190C would have had an armament of two MG 131 guns in the fuselage, four MG 151s in the wings and a fifth MG 151 mounted in the DB 603 engine to fire through the propeller spinner, with additional provision for two MK 103s or MK 108s under the wings. This, combined with the high speed and good altitude performance demonstrated at the Daimler-Benz flight test establishment, would have made the Fw 190C a singularly potent fighter had it become available, as hoped, before the end of 1943. However, the exhaust-driven turbosupercharger, upon which it depended, was a continual source of difficulty and the pressure cabin clearly required much additional development before it could be considered satisfactory for service use, and in view of the opposition to the DB 603 engine within the *Technische Amt*, Focke-Wulf eventually realised that further efforts to cure the Fw 190C of its problems would be fruitless, and all efforts were switched to the Fw 190D, which had been proceeding since mid-1942 in parallel with the former.

The Dora Story

While the Focke-Wulf design team and the powerplant engineers were struggling with the problems of pressurisation and supercharging in the Fw 190B and Fw 190C prototypes, as related in the previous chapters, work was proceeding somewhat more

The Fw 190 V20, *Werk-Nr* 0042, was first fitted with a Jumo 213A engine but later served as a prototype for the Ta 152C when a DB 603E was substituted

smoothly with the third of the trio of new variants, the Fw 190D, which, at the behest of the *Technische Amt,* was designed to take the Jumo 213 liquid-cooled Vee in-line engine. Rated at 1,776 hp for take-off in its initial production form as the Jumo 213A–1, with 1,600 hp available at 18,000 ft (5 486 m), the Jumo 213 was an engine of considerable potential, capable of being boosted by means of methanol-water injection or a two-stage supercharger, and although Kurt Tank consistently believed that the DB 603 held even greater promise for high-altitude operations, he was already committed to using the Jumo 213 in a more radical development of the Fw 190 that would emerge in due course as the Ta 152 (see p 79). Indeed, the prospective introduction of the latter influenced Tank's attitude towards the Fw 190D, which he regarded as no more than an interim solution – one that was destined, however, to serve Luftwaffe pilots well, whereas Tank's ultimate fighter would not in the end be given time to prove itself in action.

Production versions of the Fw 190D were projected on the same basis as the Fw 190C – that is to say, a start would be made with the unpressurised Fw 190D–1 and the definitive Fw 190D–2 would be pressurised. These variants, it was planned, would take advantage of the new features designed for the Fw 190 V19 (see p 68); it was intended that six more prototypes of similar standard should have been flown with DB 603 engines as part of the Fw 190C programme, but they were converted during construction to take the Jumo engine instead. These additional prototypes were the Fw 190 V20 (*Werk-Nr* 0042), V21 (0043), V25 (0050) and V28 (0053) without pressure cabin and Fw 190 V26 (*Werk-Nr* 0051) and V27 (0052) with pressure cabin. Before the decision was taken to fit these aircraft with the Junkers engine, however, three other prototypes had been put in hand to evaluate the Jumo 213 installation. These were all Fw 190A–0 airframes, being *Werk-Nummern* 0039, 0044 and 0045, redesignated as Fw 190 V17, V22 and V23 respectively. Subsequently, the Fw 190 V46 was added to the programme, being similar to the other three.

Flying by September 1942, the Fw 190 V17 had a pressurised cockpit, as did the next two prototypes, since it was still intended at that stage to incorporate this feature

The Focke-Wulf Fw 190 V53, *Werk-Nr* 170003, was one of a number of Fw 190A–8 airframes converted during 1943 to have Jumo 213A–1 engines to permit *Luftwaffe* evaluations of the proposed Fw 190D–9 production model

in the Fw 190D-2 production model. The later prototypes also had provision for a 20-mm MG 151 *Motorkannon* firing through the propeller spinner, supplementing the standard armament of two MG 131s in the fuselage and two or four MG 151s in the wings. The long-span wing used on one of the Fw 190B-0 prototypes was flown for a time on the Fw 190 V17, but was not used on any of the other prototypes or on the production versions eventually built.

The Jumo 213 installation, with an annular cowling similar to that of the Fw 190C, increased the length of the nose by 2 ft (60 cm), this being compensated for by the extra, untapered, fuselage section incorporated just ahead of the tailplane, as evolved for the Fw 190 V19. The radiators were in two semi-circular sections around the reduction gear casing, and the gills to the rear of the short cowling round the radiators operated automatically in response to a thermostat actuated by radiator temperature, to increase or decrease the flow of cooling air.

By the later months of 1943 the Jumo 213 installation in the three prototypes had been thoroughly proven, but the production plans for the Fw 190D-1 and D-2 had been shelved because no adequate cabin pressurisation system was available. Hopes that the Fw 190D would meet the high-altitude requirement of the Luftwaffe had therefore been dashed, but for operations up to 20,000 ft (6 096 m) or so, the Jumo 213-engined version still offered a substantial improvement in performance over that of the Fw 190A, and it was therefore decided to proceed as rapidly as possible with a production version based on the Fw 190A-8 airframe with the new engine installation. The original prototype was brought up to the new standard as the Fw 190 V17/U1 by June 1944. Two examples of the Fw 190A-8 were assigned for conversion to prototypes of what was now designated the Fw 190D-9 as the first production version of the *Langnasen-Dora* or, as it later became known, the *Dora–9*. These two prototypes were designated the Fw 190 V53 (*Werk-Nr* 170003) and the Fw 190 V54 (*Werk-Nr* 174024), and were in flight-test by mid-1944, although progress was delayed when V54 was severely damaged, and V53 slightly damaged, in an Allied bombing raid on 5 August 1944 on Langenhagen.

Another view of the Fw 190 V53 (see opposite). The armament shown here is similar to that fitted to the Fw 190A-8 from which the aircraft was converted, comprising two MG 151s in the upper front fuselage and four similar weapons in the wings

The Fw 190 V53 (see also pages 70–71) with the outer wing cannon removed. In this view, the 50-cm (1 ft 7½ in) extension in the rear fuselage, just ahead of the fin, can be clearly seen

The designation D–9 appears to have been chosen as a logical continuation from the Fw 190A–8; although the designation Fw 190A–9 had previously been used in connection with a few prototypes of what was intended to have been a production version, and the Fw 190A–10 had also been projected, neither of these two designations was current in 1944. An Fw 190D–8/R3 version was also projected as a ground support fighter, with two 30-mm MK 103 guns under the wings, but was not produced. Production of the Fw 190D–9 was entrusted to Focke-Wulf at Cottbus and to Fieseler at Kassel-Waldau, and deliveries to the Luftwaffe began in August 1944. Powered by the Jumo 213A–1 engine, the Fw 190D–9 had an armament of two MG 151 cannon in the wing roots, with 250 rpg, and two MG 131 guns with 475 rpg. Fuel was carried in two tanks beneath the cockpit, with a total capacity of 523 litres (115 Imp gal), supplemented by one of 115 litres (25 Imp gal) in the fuselage behind the cockpit if the MW 50 system was not fitted. Eight oxygen bottles were located in the extra parallel-section of the rear fuselage, helping to balance the added weight of the new engine in the nose.

The MW 50 bottle, when carried, was large enough to provide for 40 minutes' operation, limited to 10 minutes continuous use at a time, at altitudes up to 16,500 ft (5 030 m), but not for take-off. Production deliveries of the MW 50 were unreliable, and few Fw 190D–9s delivered before January 1945 in fact had an operational installation.

Some *Dora*–9s were fitted with *Rüstsätze* 1 or 2, respectively comprising two MG 151s or two MK 180s in the outer wings, and were desigiated Fw 190D–9/R1 and Fw 190D–9/R2; all had provision to carry two Wfr Gr 21 launchers under the wings. A few production aircraft were also equipped as all-weather fighters with the addition of *Rüstsätze*, comprising FuG 125 *Hermine* D/F navigation equipment, PKS 12 autopilot and heated windows, the designation becoming Fw 190D–9/R11. The Fw 190D–9/R14, like the Fw 190F–8/R14, was designed to carry an LT 1B torpedo or a BT 1400 bomb-torpedo on an ETC 504 rack under the fuselage. Most production Fw 190Ds had a one-piece blown canopy similar to that introduced on the Fw 190F.

The Fw 190D–10 variant was to feature a *Motorkannon* firing through the propeller shaft, as first tested on the Fw 190 V23. In this case, however, the gun in question was the 30-mm MK 108, its installation necessitating the use of the VS 19 propeller in place of the VS 9, which did not have provision for gun firing. The only other fuselage armament carried by the Fw 190D–10 was to be a single 20-mm MG 151 in the starboard side of the fuselage front decking, but no examples of this version were in fact delivered.

Another armament change distinguished the Fw 190D–11, ordered into production in January 1945, which had two MG 151s in the wing roots and two MK 108s in the outer wing bays, but without the *Motorkannon*. The engine was changed to the Jumo 213E(F) – or 9-8213H – which had a three-stage supercharger and no intercooler, with water-methanol injection before the third stage, plus a take-off rating of 1,750 hp and an emergency maximum of 2,050 hp. A

73

An early production Fw 190D-9, *Werk-Nr* 21 0051, photographed in mid-1944. A 300-l (66-Imp gal) fuel tank is carried beneath the fuselage and an early-type cockpit canopy is fitted. Below left, a three-view drawing of the Fw 190D-9 with additional side view (bottom) of the Fw 190D-0

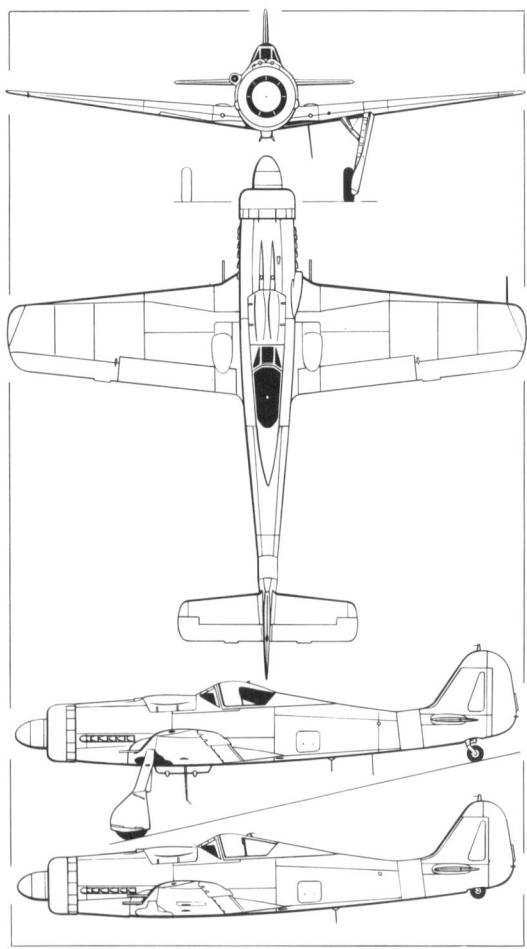

trial installation of this engine was made in a converted Fw 190A-8, which thus became the prototype of the Fw 190D-11 with the designation Fw 190 V55, making its first flight on 31 August 1944. Six more converted A-8 airframes followed, to the same specification, before the end of 1944, these having the designations Fw 190 V56 to 61 inclusive.

The Fw 190D-11 had been intended to combine the rôles of interceptor-fighter and *Schlachtflugzeugt* or ground-attack aircraft, and this theme was continued with the Fw 190D-12, which replaced the D-9 on the production lines in February 1945. This variant made use of the Jumo 213F-1 engine, with MW 50 injection, and the armament comprised one 30-mm MK 108 firing through the spinner and two 20-mm MG 151s in the wing roots. The few examples built were in Fw 190D-12/R11 all-weather fighter form, coming from the Fieseler and Arado production lines from March 1945 onwards. Two prototypes were modified from Fw 190A-8 airframes and were tested in November and December 1944, respectively, as the Fw 190 V63 and Fw 190 V64. Another prototype, the Fw 190 V65 (also a converted A–8), included the all-weather equipment and MW 50, but had additional fuel capacity in the form of four wing tanks with a total capacity of 69 Imp gal (315 litres), and was intended to set the standard for the Fw 190D–12/R5, which did not materialise.

Two other projected versions were the Fw 190D–12/R25, which would have had the Jumo 213EB engine with increased compression ratio, and the water-methanol tank in the port wing, making space for an additional 28.5 Imp gal (130-litre) fuel tank in the fuselage. Late in 1944 two prototypes of another intended production version, the Fw 190D–13/R11, were completed as the Fw 190 V62 and V71, being converted A–8 airframes. The principal change was installation of a 20-mm MG 151 in place of the 30-mm MK 108, with 220 rounds; in other respects this variant would have been similar to the Fw 190D–12/R11. The Fw 190D–12/R21, minus all-weather equipment, remained a project.

No more production versions of the *Dora* appeared, but two more variants were designated in the autumn of 1944, when the *Technische Amt* performed one of the *voltes-faces* that so bedevilled Luftwaffe plans and operations in the later stages of the war, and ordered Focke-Wulf to switch from the Jumo 213 to the DB 603 engine at the earliest opportunity. Welcoming the opportunity to vindicate Kurt Tank's consistent support for the DB 603, Focke-Wulf quickly adapted an early production Fw 190D–9 and an Fw 190D–12 to have the DB 603 engine, as the Fw 190 V76 (DB 603LA) and V77 (DB 603E) respectively, these being delivered to the Daimler-Benz *Erprobungsstelle* at Echterdingen, where their further development became the responsibility of Dr Haspel.

Demonstrating a top speed of 435 mph (700 km/h) and a service ceiling of 32,800 ft (10 000 m), these two prototypes aroused considerable enthusiasm for what became

Fw 190D–9 Specification

Power Plant: One Junkers Jumo 213A–1 12-cylinder liquid-cooled Vee engine rated at 1,776 hp for take-off and 1,600 hp at 18,045 ft (5 500 m), and with MW 50 methanol-water injection 2,240 hp at sea level and 2,000 hp at 11,150 ft (3 400 m).

Performance: (at typical loaded weight) Max speed, 357 mph (574 km/h) at sea level, 397 mph (639 km/h) at 10,830 ft (3 300 m), 426 mph (686 km/h) at 21,650 ft (6 600 m), 397 mph (639 km/h) at 32,810 ft (10 000 m); time to climb to 6,560 ft (2,000 m), 2.1 min, to 13,120 ft (4 000 m), 4.5 min, to 19,685 ft (6 000 m), 7.1 min to 32,810 ft (10 000 m), 16.8 min; max range (clean), 520 mls (837 km) at 18,500 ft (5 640 m).

Weights: Empty, 7,694 lb (3 490 kg); normal loaded, 9,480 lb (4 300 kg); max take-off, 10,670 lb (4 840 kg).

Dimensions: Span, 34 ft $5\frac{1}{2}$ in (10,506 m); length, 33 ft $5\frac{1}{4}$ in (10 192 m); height, 11 ft $0\frac{1}{4}$ in (3,36 m); wing area, 196.98 sq ft (18,3 m²).

Armament: Two 20-mm MG 151 cannon with 250 rpg in fuselage and two 13-mm MG 131 machine guns with 475 rpg in wing roots. One fuselage centreline bomb rack (ETC 504) for one 1,100-lb (500-kg) bomb.

---

known as the *Haspel–Jäger*, and plans were quickly made to modify a batch of Fw 190A–8s (and, it is believed, the Fw 190A–9 prototypes) to take the DB 603E or DB 603LA engine. This was to have been an assembly-line conversion, an initial batch of fifteen being put in hand at Echterdingen with the DB 603E engine, which was rated at 1,900 hp for take-off and 1,560 hp at 24,200 ft (7 376 m). These were to be designated Fw 190D–15, but only one had been completed by the time the Daimler-Benz plant was overrun by advancing US forces, and no examples of the Fw 190D–14, which was to have been the parallel new-production version, emerged at all.

75

Above and below left, a late-production Fw 190D-9, with the one-piece blown canopy. This example was captured in Germany during the Allied advance and was shipped to the USA for testing, as indicated by the FE-121 marking and non-standard national insignia

Production of the Fw 190D in the space of the last 12 months or so of the war totalled a little short of 700 examples, and it was operational (with III/JG 54) from October 1944 onwards at Oldenburg and, later, from Achmer and Hesepe. Despite some early misgivings among the squadron pilots, who had perhaps taken too literally Tank's own description of the *Dora-9* as no more than a stop-gap solution, the new version was soon found to offer distinct advantages over the types it replaced, whether they emanated from the Focke-Wulf stable or that of Messerschmitt. The rate of climb, the level speeds and performance in a dive were all enhanced, and the Fw 190D-9 flown by an experienced pilot could outmanoeuvre all other German fighters and most of the opposition too. By general consensus the Fw 190D proved, in the few months left to it, to be the best piston-engined fighter flown by the Luftwaffe during World War II.

Including the BMW-engined versions, overall production of the Fw 190 series totalled more than 19,000, of which 11,411 were accepted by the Luftwaffe in 1944 alone – almost four times the total for the previous year. Despite severe damage suffered by many of the manufacturing units, output was maintained at a quite remarkable level almost up to the moment that each factory was occupied by advancing Allied troops – although the merits of doing so for a fuel-starved Luftwaffe might well be argued.

76

Metamorphosis

The attempts to improve the high-altitude performance of the Fw 190 – undertaken largely on an *ad hoc* basis – have already been described, and it has been shown that these efforts failed in their primary purpose, although they did lead to the production of improved models of the Fw 190. During 1942, the need for high-altitude interceptors was pressed in upon the Luftwaffe by the mounting tempo of USAAF activities in Europe, coupled with intelligence reports of the development of new bomber types with pressure cabins in both the UK and the USA. Before the end of the year, consequently, an outline specification was drawn up for a new high-performance fighter *(Hochleistungsjäger)* in which emphasis was to be placed upon the high-altitude performance with provision also to be made for its operation in the reconnaissance-fighter rôle at medium-to-high altitudes. A two-stage programme was suggested, using in the short term *(Sofort-Programm)* an adaptation of an existing production fighter, followed in the longer term *(Vorrucken-Programm)* by a completely new design.

Focke-Wulf and Messerschmitt were invited to submit proposals for both phases of the *Hochleistungsjäger* and both, in due course, were to receive the backing of the *Technische Amt.* For the short-term answer, both companies naturally proposed variants of their respective contemporary fighters with long-span wings to increase the combat

The Focke-Wulf Ta 152 V7, *Werk-Nr* 11 0007 which served as a prototype for the Ta 152C–0/R11, with all-weather equipment. This prototype was also earmarked to be used for torpedo-dropping trials, in which case it would have represented the Ta 152C–0/R14 variant

Above and below left, two views of the Fw 190 V32/U1, with DB 603G engine and a long-span wing. Later, this same airframe became a prototype for the Ta 152H as the Fw 190 V32/U2, with Jumo 213 engine and a still longer-span wing

ceiling and other modest changes; for the long-term solution, Messerschmitt offered a totally new design that was to emerge eventually as the Me (later BV) 155B, while Focke-Wulf based its proposal on the Fw 190 with an extensive structural redesign and various new aerodynamic features.

Between 1940 and 1942, the Focke-Wulf design team under the direction of Dipl-Ing Kurt Tank had undertaken numerous project studies based on the Fw 190, particularly in respect of various engine installations, armament changes and wing redesign. Among the engine possibilites studied, in addition to later versions of the BMW 801 itself, were the BMW 802 and BMW 803 radials, the DB 609, DB 614, DB 623 and DB 624 inline units and the Junkers Jumo 222 inline engine. Most of these project studies were conducted under Fw 190V designations, in some cases conflicting with the *Versuchs* numbers eventually used on prototypes that were completed and test-flown. Another series of designations came into use for advanced projects of the fighter during 1942, commencing with the Fw 190Ra–1, and these were mostly concerned with improving the high-altitude performance. The Ra–1 and Ra–4 were based upon the Jumo 213 engine, as already selected for installation in the Fw 190D, whereas the Ra–2 and Ra–3 were designed around the BMW 801D, and the Ra–6 was projected with either a Jumo 213E, a DB 603G or DB 632. Various new features were designed for these projects, some having a high aspect ratio wing with a span of 48 ft $6\frac{1}{2}$ in (14,8 m) and area of 242.2 sq ft (22,5 m²), and all featuring a new lengthened fuselage with the cockpit relocated 16 in (40 m) farther aft in relation to the wing, and the vertical tail surfaces enlarged. A new centre section was designed, so that those projects retaining the basic Fw 190 wing, rather than having the new long-span version, still had an increase in overall span, to 36 ft 1 in (11,00 m), the area going up to 209.89 sq ft (19,5 m²).

When the *Technische Amt* accepted, in 1943, the basic Focke-Wulf proposals for development of the fighter along these lines, Tank was accorded the unusual honour of having the new type designated after him, instead of it being given a new 'Fw' designation. Although several of the official designation groups in use already related to the name of an individual – eg, Ju for Junkers, Me for Messerschmitt – they had been assigned on the basis of *company* names, and the use of Ta for aircraft that were still to be responsibility of the Focke-Wulf company was an indication of the high esteem attained by Tank by this time (the precedent had only one equal, when Dipl-Ing Kalkert's work for Gothaer Waggonfabrik was recognised in the designation Ka 430).

The first Tank designation to be assigned was Ta 152, intended to cover the development of the Fw 190 along the lines of the Ra–1 and Ra–4 proposals with a Jumo 213 engine, various structural improvements, heavier armament and, in some models, pressure cabin and long-span wing. The designation Ta 153 seems to have been assigned to a further variant using a DB 603 engine or one of its later derivatives such as the DB 623 or DB 627, but no examples of this were completed and its exact specification remains obscure. Work on the Ta 152 also proceeded extremely slowly and although Tank himself had high hopes for it and, as already shown, regarded the similarly powered Fw 190D as no more than a stopgap, the newer type was never to reach full operational status and its full fighting potential thus remained unrealised.

The designation of the sub-series of Ta 152s has been the cause of some confusion, since some effort appears to have been made to integrate certain of the variant designations with those of the Fw 190 family, while others were issued in a logical sequence order. The initial Ta 152A–1 version was projected as a medium-altitude *Schwerstjäger* (heavy fighter – a reference to its armament) with a Jumo 213A mounting a 30-mm MK 103 *Motorkannon*, plus two 20-mm MG 151s in the fuselage and one similar gun in each wing root, and two 30-mm MK 108s (or, in the Ta 152A–2, 20-mm MG 151s) in the outer wing bays. However, these versions were abandoned before prototypes were built and the designation Ta 152B was adopted instead, apparently at the same time that the designation of the engine modified to accept a hollow shaft for the gun installation was changed to Jumo 213C.

The Ta 152B had the new fuselage, based on that of the Fw 190A–8 but with an extra section added in front of the cockpit and the 1 ft 7½ in (50 cm) section in the rear fuselage, as adopted for the Fw 190D. The new wing centre section was used, as already noted,

The Fw 190 V30/U1, originally one of the Fw 190C experimentals, was later rebuilt in this form as a prototype for the Ta 152H with Jumo 213 powerplant and long span wing

79

Focke-Wulf Ta 152H-1 Cutaway Drawing Key:
1 Starboard navigation light
2 Pitot tube
3 Wing skinning
4 Aileron tab control linkage
5 Aileron tab
6 Starboard aileron
7 Aileron controls
8 Pitot tube heating
9 Wing lateral stringers
10 Flap controls
11 Flap panels
12 Flap actuating jack
13 Starboard wing fuel tanks (three bag-type)
14 Undercarriage indicator
15 Abbreviated steel front spar
16 Auxiliary intake
17 Supercharger air intake
18 Cooling louvres
19 Junkers three-blade wooden propeller
20 Spinner
21 Cannon port
22 Blast tube
23 Annular radiator
24 15-mm ring armour
25 Cooling gills
26 Starboard mainwheel
27 Exhaust stubs
28 Anti-vibration mounting pads
29 30-mm MK 108 cannon
30 Forged engine bearer
31 Engine accessories
32 Supercharger inlet trunk
33 Junkers Jumo 213E engine
34 Generator
35 No 1 fuselage frame
36 Oil tank, capacity 15.8 Imp gal (72 l)
37 Engine bearer/bulkhead attachment
38 Firewall
39 Engine bearer support member
40 Cannon shell ejector chute
41 Front spar carry-through
42 Front spar/fuselage attachment
43 Cannon ammunition box (90 rounds)
44 Cockpit forward pressure bulkhead (No 1A fuselage frame)
45 Cannon retardation/resistance mechanism
46 Instrument panel
47 Gunsight mounting
48 Control column
49 Rudder pedals
50 Underfloor control linkage
51 Floor support members
52 Cockpit floor (armoured)
53 Seat harness attachment
54 Pilot's seat (armoured)
55 Instrument panel shroud
56 Revi 16B gunsight
57 Armoured-glass windscreen
58 Starboard instrument console
59 Canopy rubber-tube pressurization
60 Rearward-sliding cockpit canopy
61 Headrest
62 20-mm head-armour
63 Turn-over bar and shroud
64 5-mm shoulder-armour (two-piece)
65 8-mm back-armour (two-piece)
66 Lead storage battery
67 Cut-out box
68 Dynamo
69 Cockpit rear pressure bulkhead (No 8 fuselage frame)
70 FuG 125 navigation equipment (only in H-1/R11 all-weather variant)
71 Distributor
72 GM 1 tank, capacity 18.7 Imp gal (85 l)
73 Tank armour (attached to No 9 fuselage frame)
74 Radio bay access hatch
75 LGW-Siemens K 23 autopilot
76 FuG 16ZY radio transmitter/receiver
77 No 10 fuselage frame
78 Rudder control rod
79 Compressed air line
80 Master compass
81 Elevator control cables
82 No 12 fuselage frame
83 Fuselage construction
84 AZA 10 signal cartridges (port and starboard)
85 Lift/hoist tube
86 Rudder rod/cable transition
87 Aerial lead-in and adaptor
88 Aerial
89 Cylindrical fuselage extension (frame Nos 14-16)
90 Oxygen cylinder stowage shelf
91 Compressed air bottle for cannon operation, capacity 1.1 Imp gal (5 l)
92 Elevator control quadrant
93 Fuselage/fin joint
94 Starboard tailplane
95 Elevator balance
96 Starboard elevator
97 Elevator tab
98 Fin construction
99 Tailwheel retraction cable
100 Rudder upper hinge
101 Rudder construction
102 Tailwheel leg retraction guide
103 Rudder hinge control
104 Rudder tab
105 Rear navigation light
106 Electric lead
107 Elevator torque tube
108 Tailwheel shock-absorber
109 Elevator tab
110 Elevator balance
111 Elevator construction
112 Semi-retractable tailwheel (380 x 150-mm)

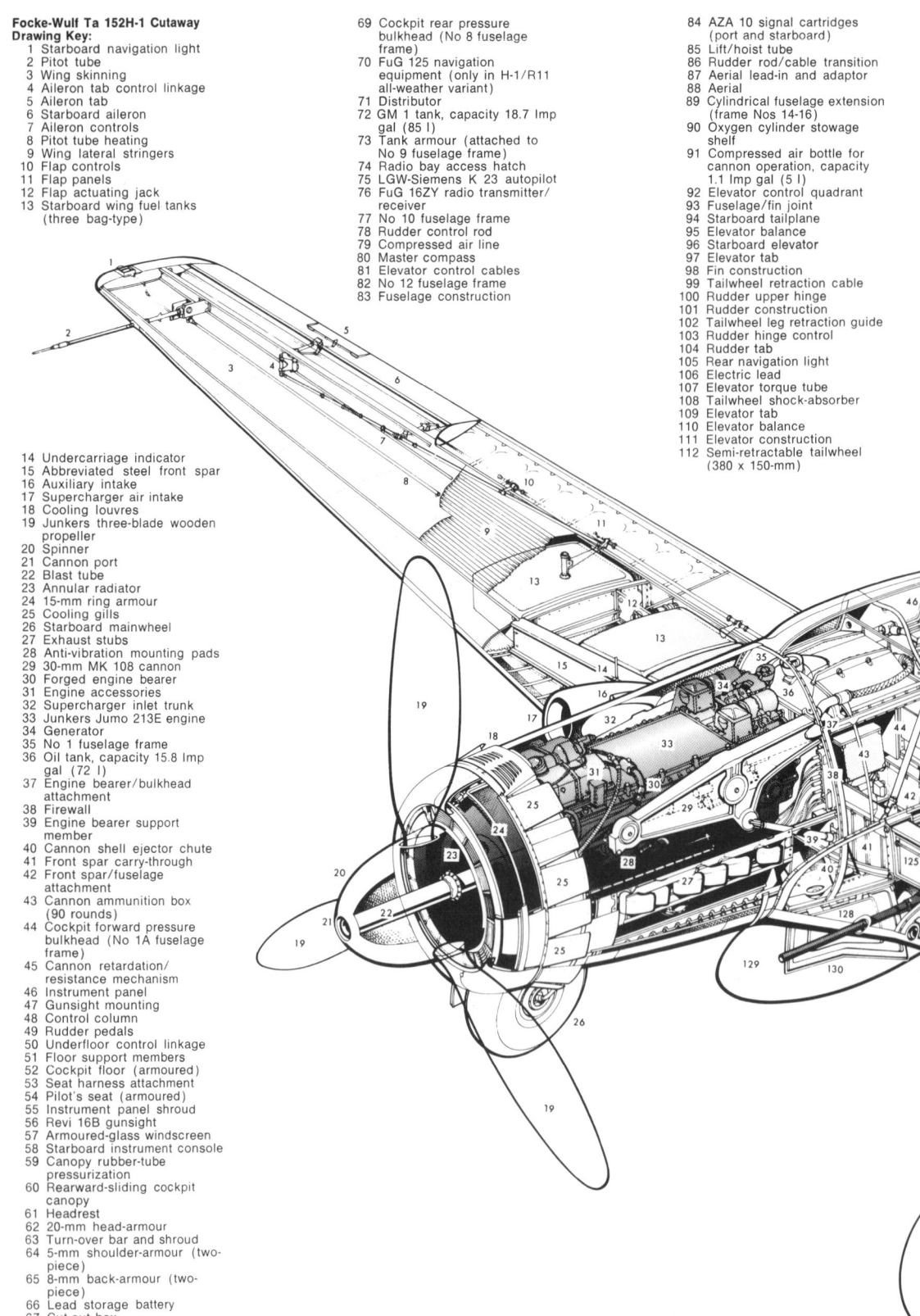

113 Fin spar attachment
114 Antenna
115 D/F loop
116 Retractable entry step
117 Spring-loaded hand/foothold
118 Rear fuselage fuel tank (protected), capacity 80 Imp gal (362 l)
119 Rear spar/fuselage attachment
120 Forward fuselage fuel tank (protected), capacity 51 Imp gal (233 l)
121 Wing gun breech fairing
122 Port MG 151/20 wing gun
123 Shell ejector chute
124 Port ammunition box (175 rounds)
125 Undercarriage retraction guide track
126 Wing gun forward mounting
127 BSK 16 camera gun
128 Gun barrel
129 Auxiliary drop tank, capacity 66 Imp gal (300 l)
130 Port inboard undercarriage door
131 Ventral antenna
132 Undercarriage retraction strut
133 Towing lug (port and starboard legs)
134 Undercarriage leg
135 Port mainwheel (740 x 210-mm)
136 Brake cable
137 Axle
138 Port outboard undercarriage door
139 Shock-absorbers
140 Mainwheel leg fairing
141 Mainwheel leg pivot point

142 Undercarriage indicator
143 Abbreviated steel front spar
144 Fuel pump
145 VHF interference suppressor
146 Port inboard wing tank (MW 50), capacity 15.4 Imp gal (70 l)
147 Port navigation light electric lead
148 Port centre wing tank (B4 fuel)
149 Mainwheel leg attachment plate (spar rear face)
150 Flap actuating jack
151 Port outboard wing tank (B4 fuel)
152 Flap structure
153 Wing lateral stringers
154 Wing rib stations
155 Wing skinning
156 Aileron tab control linkage
157 Aileron tab
158 Port aileron
159 Full-span rear spar
160 Port navigation light

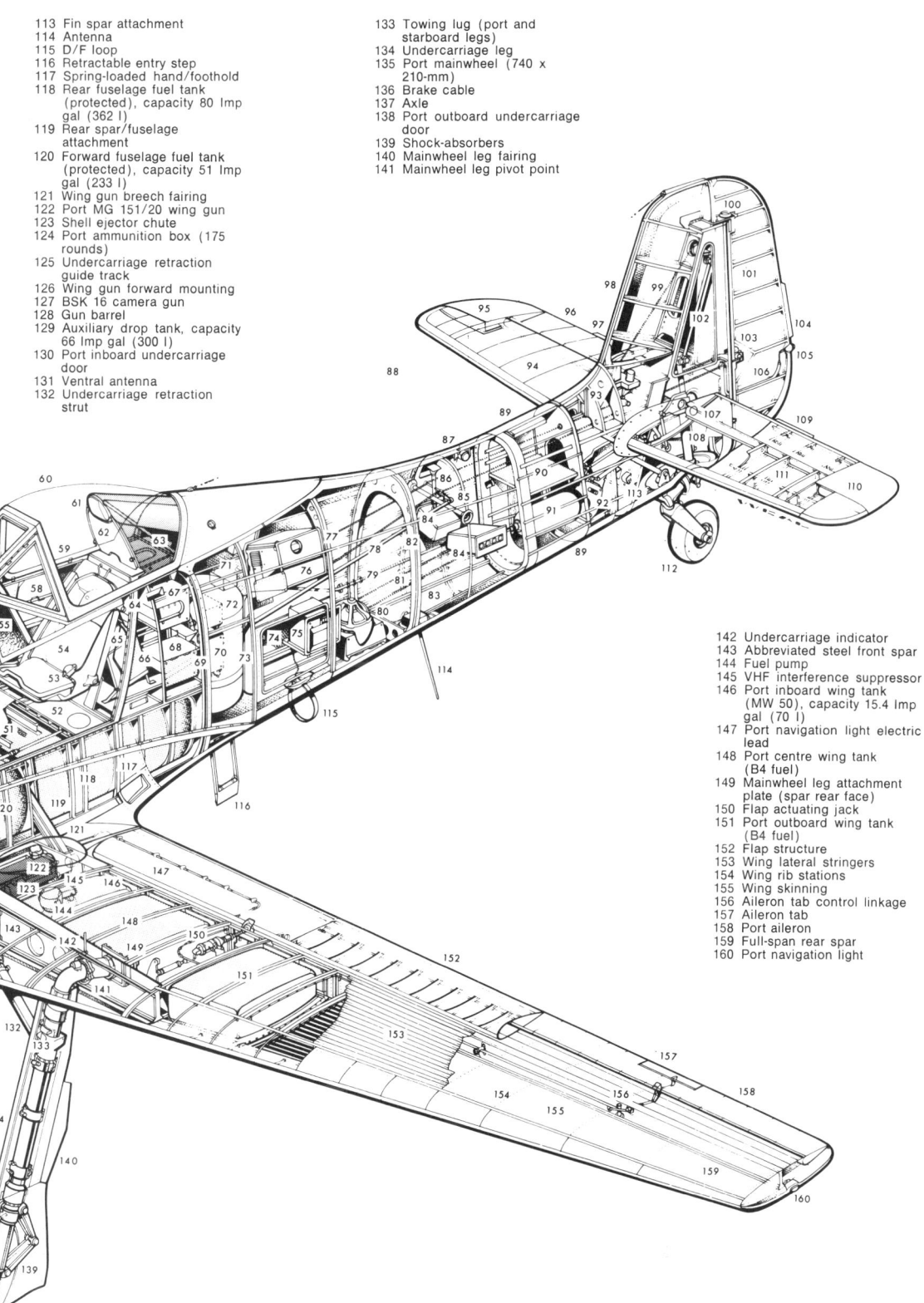

81

increasing the span to 36 ft 1 in (11,00 m) and widening the undercarriage track also, since the main leg attachment points remained unchanged in the standard Fw 190 outer wing panels.

While plans were being made to produce the Ta 152B, a second version was put in hand to meet the Luftwaffe's still outstanding requirement for a high-altitude fighter. This new version was designated Ta 152H, having the Jumo 213E engine with an MK 108 *Motorkannon* plus one MG 151 in each wing root, the same fuselage as the Ta 152B, a pressurised cockpit and the high aspect ratio wing of 47 ft 4½ in (14,44 m) span. The 'H' designation appears to have derived from the earlier intention to develop an Fw 190H high-altitude fighter based on the Fw 190Ra-6 project, with a new wing married to a lengthened version of the Fw 190B fuselage.* In any case, there seems to be no substance for a theory that the Ta 152B and Ta 152H designations indicated, respectively, *Begleitjäger* (escort fighter) and *Hohenjäger* (altitude fighter), since the Ta 152H was itself referred to as a *Begleitjäger* when it was assigned the task of flying high cover for Me 262 operations; such a theory also would not account for the Ta 152C and Ta 152E designations, which were in use at about the same time.

To serve as prototypes for the Ta 152 series, several of the experimental Fw 190s were allocated for conversion, while contracts were placed with Focke-Wulf for a pre-production batch of Ta 152 airframes to be built at Sorau, and preparations were made for full-scale production at Cottbus. The Sorau-built aircraft received the *Werk-Nummern* 11 0001 to 11 0026 inclusive, and all were destined to be assigned *Versuchs*

---

* An alternative theory regarding Ta 152 designations is that Ta 152A was abandoned to avoid confusion with Fw 190A, and that Ta 152D, F and G were omitted for the same reasons. This would make a logical progression of Ta 152B, C, E and H, but does not explain how the Ta 152H apparently predates the Ta 152C and E.

Ta 152B-5/R11 Specification

Power Plant: One Junkers Jumo 213E-1 12-cylinder liquid-cooled Vee in-line engine rated at 1,750 hp for take-off and 1,320 hp at 32,810 ft (10 000 m) or 1,740 hp with GM 1 nitrous oxide boost.

Performance: Max speed, 342 mph (550 km/h) at sea level, 428 mph (689 km/h) at 35,270 ft (10 750 m), and (with GM 1) 443 mph (713 km/h) at 44,290 ft (13,500 m); max continuous cruising speed, 315 mph (507 km/h) at 24,605 ft (7 500 m); range (clean, at max continuous cruise speed) 775 mls (1 247 km) at 385 mph (620 km/h) at 29,530 ft (9 000 m); range (with 66 Imp gal/300-litre drop tanks, at economical cruise), 1,180 mls (1 900 km) at 300 mph (483 km/h) at 21,325 ft (6 500 m).

Weights: Normal loaded, 10,750 lb (4 875 kg); max take-off, 11,900 lb (5 400 kg).

Dimensions: Span, 36 ft 1 in (11,00 m); length, 35 ft 1½ in (10,710 m); height, 11 ft 0¼ in (3,36 m); wing area, 209.89 sq ft (19.5 m²).

Armament: One 30-mm MK 103 cannon in engine, firing through spinner and one MK 103 in each inner wing, with 80 rpg. Provision for one fuselage centreline bomb rack (ETC 503) with capacity up to 1,100 lb (500 kg).

---

numbers, as Ta 152 V1 to Ta 152 V26 inclusive. The production aircraft from Cottbus were numbered 15 0001 and up, and two of these became the Ta 152 V27 and Ta 152 V28.

The first Ta 152s appeared at Langenhagen in mid-1944, these being the two initial aircraft assembled at Sorau to Ta 152H standard, and the second of these was transferred to Rechlin in August. Almost simultaneously, the Fw 190 V33/U1 appeared at Langenhagen, this being one of the Fw 190C prototypes rebuilt with Jumo 213E-1 engine. The wing of the Ta 152H had an unusual structure, the all-steel front spar extending only just beyond the undercarriage attachment points. The rear spar was full-span in length and the wing derived its strength and rigidity from close-set ribs and numerous lateral stringers to reinforce the stressed skin. This wing was used on the Ta 152 V1 and Ta 152 V2 prototypes from the outset, but the Fw 190 V33/U1 introduced the definitive fuel tankage for the Ta 152H–1, comprising three tanks in each inner portion, located just aft of the truncated mainspar. The inner port tank contained 15.4 Imp gal (70 litres) of methanol water for the MW 50 installation, which boosted take-off output of the Jumo engine from 1,750 hp to 2,050 hp, and provided 1,800 hp at 26,250 ft (8 000 m) with 485 lb (220 kg) of residual exhaust thrust also available at that altitude. The remaining five wing tanks had a capacity of 88.5 Imp gal (400 litres), supplementing the main fuselage tank of 130 Imp gal (600 litres). Behind the fuselage tank there was an 18.7 Imp gal (70 litre) tank of nitrous oxide for the GM 1 system, which was capable of boosting the engine output at 32,810 ft (10 000 m) from 1,320 hp to 1,740 hp.

First flown on 12 July 1944, the Fw 190 V33/U1 was unfortunately lost on its second flight on the next day, setting back the flight test programme. Three other former Fw 190C prototypes, all of which like the Fw 190 V33 had originally had exhaust-driven turbosuperchargers in ventral housings, had been assigned to the Ta 152 programme, and the next of these to appear was the Fw 190 V30/U1, first flown in its new guise on 6 August 1944. Whereas the Fw 190 V33 had had a Jumo 213E–1, V30 had the Fw 190D's standard Jumo 213A–1 power egg, and did not have the new wing tanks; this prototype also was ill-fated and survived only a week of flight testing before crashing on 23 July 1944.

The Fw 190 V29/U1, completed by the end of September, had a Jumo 213F engine with a three-stage supercharger rated at 2,060 hp, and was soon followed by the Fw 190 V32/U2 (this airframe having served Daimler-Benz as an engine test-bed with the designation Fw 190 V32/U1 in the interim), which received a set of Ta 152H–0 wings originally destined for the Ta 152 V25, construction of which was abandoned. Also committed to the test programme at this time was the Fw 190 V18/U2, a second rebuild of the old Fw 190A–0 *Werk-Nr* 0040, but this aircraft, too, crashed within a few days of its first flight in the new form, on 8 October 1944. Three more of the pre-production Ta 152s were at Langenhagen by the end of the year, as the Ta 152 V3, V4 and V5, these being in Ta 152H–1 configuration.

As already noted, the Ta 152H was designed to carry an armament of one 30-mm MK 108 *Motorkannon* with 90

A three-view drawing of the Ta 152C–0

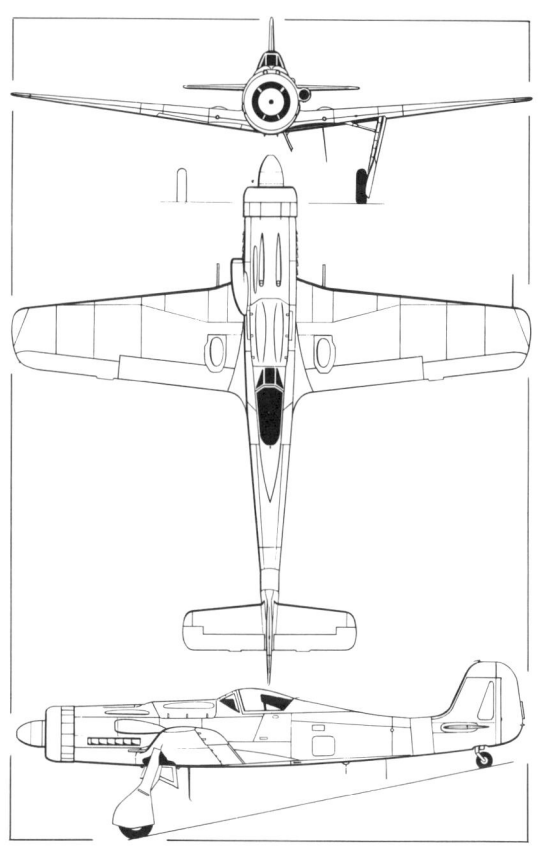

The Ta 152H–0 *Werk-Nr* 150005 – the fifth production airframe – on the compass-swing platform at Cottbus prior to delivery to the *Luftwaffe*. Of the numerous versions of the Ta 152 projected or built in prototype form, only the Ta 152H saw active service

rounds and two MG 151s in the wing roots with 175 rpg, and an ETC 503B rack on the fuselage centreline to carry a 66 Imp gal (300 litre) fuel tank. It had a pressurised cockpit, the boundaries of which were made up by the armoured forward bulkhead, the floor, the rear armour plate and cockpit sidewalls, all rendered airtight, and the enclosure was completed by the sandwich-type armourglass windscreen and sliding canopy, the latter being sealed by means of rubber packing pieces. A pressure differential of 4.4 lb/sq in (0,30 kg/cm²) could be achieved, giving an equivalent altitude of about 16,500 ft (5 000 m) when the aircraft was at 35,000 ft (10 600 m).

The initial batch of twenty aircraft off the Cottbus production line were of Ta 152H–0 type, without the extra wing tanks and without MW 50 and GM 1 systems. These went to Rechlin from October 1944 onwards for service trials in the hands of *Erprobungskommando Ta* 152 under *Hauptmann* Bruno Stolle, and this unit conducted some operations with the type. Production continued with the Ta 152H–1, with the full fuel capacity and boost systems as first tested on the Fw 190 V33/U1. The Ta 152H–2 was to have had the improved FuG 15 radio in place of the FuG 16Z–Y normally fitted, but production of this variant was cancelled in December 1944 with none completed. Also remaining in the project stage were the Ta 152H–10, H–11 and H–12, which were to have been respectively the same as the H–0, H–1 and H–2 but with provision to carry cameras in the fuselage behind the cockpit.

One Ta 152H–0 and most of the Ta 152H–1s were equipped for all-weather operation by means of a *Rüstsatz,* this taking the number R11 as it was identical with that used on the Fw 190D. Apart from their use by the experimental unit at Rechlin, some Ta 152H–1s saw action with the *Stabsschwarm* of the JG 301 fighter wing from January 1945 onwards, more than 150 examples having been completed before the Cottbus works had to be abandoned in the face of the Soviet advance. A few were assigned to the *Mistel* programme but the principal operational rôle for the Ta 152H proved to be that of flying high cover for the Messerschmitt Me 262 jet fighters, which were particularly vulnerable to attack from Allied fighters during take-off and landing.

84

The other variants of the Tank fighter, on which work proceeded during the final months of 1944 and early 1945, progressed no further than prototypes. As previously explained, the Ta 152A had given way to the Ta 152B, the B-1 and B-2 versions of which were proposed around the Jumo 213C engine, with GM 1 boost. With MG 151 cannon in the forward fuselage and wing root positions, the B-1 was to have a 30-mm MK 108 *Motorkannon* while the B-2 would have the MK 103 weapon of similar calibre. In both cases /R1 and /R2 versions were planned, respectively with two MG 151s and two MK 108s in the outer wing bays, but the Jumo 213E was then proposed in place of the 213C, and the designations Ta 152B-3 and Ta 153B-4 were assigned instead. Neither of these was destined to appear even as a prototype, however, and the Fw 190 V68 (a reconstruction of the Fw 190 V53, which had been, in effect, an Fw 190D-0) was completed late in 1944 to represent the Ta 152B-5. The difference from the earlier proposed versions lay in the armament, which was reduced to just three 30-mm MK 103s – one in the engine and one in each wing root.

Production models were to have MW 50 and the R11 all-weather kit incorporated, and were represented by three of the development batch aircraft, which were

Above and below, two views of the sole Ta 152H-1, *Werk-Nr* 150168, taken to the Royal Aircraft Establishment at Farnborough after being captured in 1945. Few flights were made in the aircraft at the RAE

With unauthentic *Luftwaffe* markings crudely applied over RAF roundels and fin flash, this captured Ta 152H–1 was taken to the USA for trials with the FE (Foreign Evaluation) number 112

designated Ta 152 V19, V20 and V21. Plans to have the Ta 152B–5/R11 ready for delivery by March 1945 had to be abandoned in the event because of the higher priority afforded the Ta 152H.

Similarly, the Ta 152E remained stillborn, this being the projected fighter-reconnaissance version of the Ta 152B. Two of the development aircraft were assigned to this programme as Ta 152 V9 and Ta 152 V14, but were never flown, eventually being used for static testing. With the same armament as the Ta 152B–5, the E model had an Rb 75/30, Rb 20/30, Rb 50/30, Rb 30/18, Rb 50/18 or similar twin-camera installation vertically mounted in the rear fuselage; projected versions were the Ta 152E–0 without MW 50 or GM 1, the Ta 152E–1 with both the boost systems, and the high-altitude Ta 152E–2. The last-named of this trio would have had the long-span wing of the Ta 152H, but was abandoned in favour of the Ta 152H–10 — itself destined to be stillborn.

When the *Technische Amt* made its belated (at least in Kurt Tank's view)

A three-view drawing of the Ta 152H–0

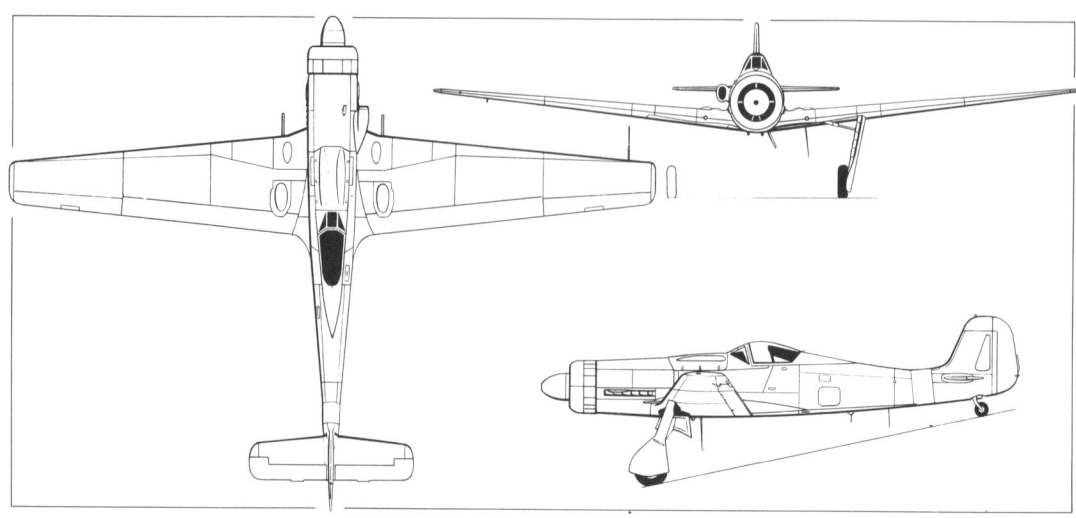

decision, in August 1944, to allow the DB 603 to be used in the Focke-Wulf fighters in place of the Jumo 213, a maximum effort was mounted to permit the Ta 152 to take advantage of the Daimler-Benz engine, in a version known as the Ta 152C. Apart from the engine installation, this was basically to be the same airframe as the Ta 152B, and while prototypes were rushed ahead, production plans were made to allow deliveries to begin in April 1945 from the ATG concern at Leipzig and Siebel at Halle. Two of the Jumo-engined Fw 190D prototypes were earmarked for conversion in August 1944, to have DB 603L engines, as the Fw 190 V20/U1 and V21/U1, but the first of these airframes was destroyed in an air raid on 5 August before the work could begin. Fw 190 V21/U1 emerged on 3 November 1944 with a DB 603E and was assigned to Daimler-Benz for engine testing, and the same engine type was installed in three of the development batch Ta 152s, which were designated Ta 152 V6, V7 and V8 and had the cowlings designed for the production Ta 152C with DB 603L. Three more development aircraft, the Ta 152 V19, V20 and V21, had been selected also for DB 603L testing but were switched to Ta 152B-5/R11 configuration in March and April 1945. Six other airframes intended as prototypes of Ta 152C variants were cancelled.

The Ta 152 V6, with a DB 603LA engine, was otherwise similar to the Fw 190 V21/U1 and made its first flight on 28 February 1945. These prototypes represented the proposed Ta 152C-0 and C-1 production variants, with an armament of one 30-mm MK 108 *Motorkannon* and four 20-mm MG 151s in the fuselage and wing roots. The Ta 152 V7 was similar, with the addition of the R11 all-weather kit, thus representing the Ta 152C-0/R11, and the Ta 152 V8 incorporated the new Revi EZ-42 gyro gunsight manufactured by Askania and with which it was hoped to equip the Ta 152C-1.

Ta 152H-1 Specification

Power Plant: One Junkers Jumo 213E-1 12-cylinder liquid-cooled Vee in-line engine rated at 1,750 hp for take-off or 2,050 hp with MW 50 methanol-water injection and 1,320 hp at 32,810 ft (10 000 m) or 1,740 hp with GM 1 nitrous oxide boost.

Performance: Max speed, 332 mph (534 km/h) at sea level or 350 mph (563 km/h) with MW 50; 465 mph (748 km/h) at 29,530 ft (9 000 m) with MW 50 and 472 mph (760 km/h) at 41,010 ft (12 500 m) with GM 1; continuous cruising speed, 311 mph (500 km/h) at 22,965 ft (7 000 m); initial rate of climb, 3,445 ft/min (17,5 m/sec) with MW 50; service ceiling, 48,550 ft (14 800 m) with GM 1; range (clean, at max continuous cruise speed), 755 mls (1 215 km) at 376 mph (605 km/h) at 32,810 ft (10 000 m); range (with 66 Imp gal/300 litre drop tank, at economical cruise), 1,250 mls (2 000 km) at 293 mph (472 km/h) at 22,965 ft (7 000 m).

Weights: Empty equipped, 8,642 lb (3 920 kg); normal loaded, 10,472 lb (4 750 kg); max take-off, 11,502 lb (5 217 kg).

Dimensions: Span, 47 ft 4½ in (14,440 m); length, 35 ft 1½ in (10,710 m); height, 11 ft 0¼ in (3,36 m); wing area, 250.8 sq ft (23,3 m²).

Armament: One 30-mm MK 108 engine-mounted cannon firing through airscrew spinner with 90 rounds and one 20-mm MG 151 cannon in each inner wing with 175 rpg.

Whereas the Ta 152 V6 and V8 flew with the DB 603L, the V7 was fitted with the DB 603EM, which offered 1,800 hp for take-off, boosted to 2,250 hp by the use of MW 50, an improvement of 150 hp over the output of the DB 603L. Comparative trials showed that the Ta 152 V7, at a gross weight of 11,706 lb (5 300 kg), had sea level maximum speeds of 342 mph (550 km/h) and 370 mph (595 km/h) without and with MW 50 injection, whereas the DB 603L-

Another view of the Ta 152H-1 that was tested in the USA – see also page 86

engined variants had speeds of 339 mph (545 km/h) and 359 mph (578 km/h) respectively. At higher altitudes, on the other hand, the DB 603L provided a substantially better performance. Since both the DB 603EM and the DB 603L used 96 octane C3 fuel, which was becoming critically short in Germany in the early months of 1945, it was eventually decided to standardise on the DB 603LA in the Ta 152C, as this unit could use C3 or the lower-rated (87-89 octane) B4 fuel.

The proposed Ta 152C-2 would have had improved radio equipment. It was abandoned, however, and the production effort was concentrated on the C-1 and the C-3 versions, the latter substituting the 30-mm MK 103 with 80 rounds for the MK 108 with 90 rounds used in the Ta 152C-1. All-weather versions of both were to be built, with the R11 *Rüstsatz,* and there was also a proposal to equip some C-1s with ETC 504 torpedo racks as Ta 152C-1/R14s. Installation of a 45.8 Imp gal (208 litre) GM 1 tank in the rear fuselage, with a counterweight in the engine bay, was proposed with the R31 kit for the C-1. Armament changes were projected for the Ta 152C-4 and Ta 152C-5 heavy fighters, the former having FuG 16Z-Y radio and Wfr Gr 21 rocket launchers under the wings; prototypes designated Ta 152 V22, V23 and V24 were abandoned. The Ta 152C-5 was to have two MK 103s in the wing roots, in place of the 20-mm weapons, and the Ta 152C-11 was a proposed reconnaissance variant using the fuselage of the Ta 152E.

Drawings were completed for two-seat training versions of the Ta 152C, with lengthened cockpit similar to that of the Fw 190A-8/U1, but no examples of these Ta 152S-1 and S-2 versions, with DB 603LA, were built. Many other projects also received some attention in the Focke-Wulf drawing offices, such as the development of a high-performance low-altitude version, but by the Spring of 1945 it had become obvious to all but the most fanatical adherents of the Third Reich that any thought of further evolution of new aircraft was pointless and that only the continued production of proven types already in service could conceivably be of any benefit to the Luftwaffe's sorely tried fighting personnel. With Soviet and US forces sweeping into Germany from north and south, however, even production facilities were soon to be disrupted and plans for production of Ta 152Cs by Fieseler and by Roland fell by the wayside, while Siebel and ATG were still at the component production stage when their assembly lines were halted.

The Butcher-Bird in Combat

If the hallmark of the thoroughbred in fighter aircraft is overall amenability to widely differing operational demands, then this was never demonstrated more forcibly than by the operational career of the Focke-Wulf Fw 190. This hallmark was also to be seen in the Messerschmitt Bf 109, the Supermarine Spitfire, the North American P–51 Mustang and the Mitsubishi A6M Zero-Sen; all these fighters were outstanding to varying degrees of excellence in their ease of pilot handling and field maintenance, their combat performance and their suitability to fill various operational rôles, but none more so than Dipl Ing Kurt Tank's Fw 190.

Entering service with II/JG 26 on the Channel Front in July 1941, the Fw 190A–1 soon established an ascendancy over all comers; if its début was marred by cooling problems associated with its BMW 801C–1 engine, it nevertheless demonstrated all the potential of an outstanding combat fighter. It could out-perform the RAF's Spitfire Mk VB on every count with the exception of

In combat, the Focke-Wulf Fw 190 proved a highly successful fighter and fighter-bomber after some early problems had been overcome. In this propaganda shot, an early example formates on an Fw 189

turning circle and, as one successful British fighter pilot commented acidly, 'Turning doesn't win battles!' It was not until the introduction of the P-51B Mustang, in December 1943, that the Allies at last fielded a fighter that could produce combat parameters over and above those of the then current Fw 190A-6. The Spitfire Mk IX with the Merlin 61 engine, which entered service with RAF squadrons from July 1942 onwards, had gone a long way towards redressing the balance and had even displayed an edge over the Fw 190 in climb and performance above 26,000 ft (7 925 m), but when in trouble, the Focke-Wulf's half-roll and dive still left the Spitfire standing.

By the summer of 1944, the overall performance of the latest BMW 801-engined Focke-Wulfs — the A-7 and A-8 — had fallen marginally behind the performances proffered by the latest Allied types, such as the Tempest, the P-51D Mustang and the Spitfire XIV, but the status quo was to be more than restored by the tardy introduction of the Fw 190D-9, which, powered by the Junkers Jumo 213A-1 liquid-cooled engine with emergency power boosting, was to be held by many Luftwaffe pilots – and not without justification – to be the best Fw 190 of them all.

In July 1941, the intensity of the RAF day offensive against targets in Northern France was peaking. Implemented by Air Vice Marshal Leigh-Mallory and referred to as the 'Non-Stop' offensive, its prime objective was that of tying down in the West as many *Jagdgruppen* as possible and thereby precluding their transfer to the Soviet front. Importance was also attached to the rapid build-up of operational experience that the intensive operations were providing the Spitfire and Hurricane squadrons of Nos 10 and 11 groups, and not to be discounted was the salutary effect on home morale through the Press media of the accounts of daily combat with the Luftwaffe over France and of bomber raids on targets in enemy-held territory. The offensive took the form of almost daily 'Circus' operations in which a number of Blenheims or Stirlings, escorted by up to 300 Spitfires and Hurricanes, penetrated enemy air space over France with the object of bringing the Luftwaffe to battle. There was no element of surprise and often the escort found itself fighting at a grave tactical disadvantage, with time being on the side of the Luftwaffe.

The German fighter units remaining in France after Operation 'Barbarossa', the invasion of the Soviet Union, on 22 June

With a bomb under the fuselage and empty wing racks, an Fw 190 gets airborne for another mission

Wing bombs and an empty fuselage rack are carried by this Fw 190G seen at the moment of undercarriage retraction

1941, were *Jagdgeschwader* 2 'Richthofen' under *Hauptmann* Wilhelm Balthasar and *Jagdgeschwader* 26 'Schlageter' under the redoubtable *Oberstleutnant* Adolf Galland. Both units were operating a mix of E–7, F–2 and F–4 subtypes of the Messerschmitt Bf 109, the last-mentioned having actually entered service during June, and with the exception of the *I Gruppe* of JG 2, which was based near Brest, all the *Jagdgruppen* came under *Jagdfliegerführer* 2 (Fighter Leader 2), with the Staff of JG 26 at Audembert, I/JG 26 at St Omer-Clairmarais, II/JG 26 at Maldeghem and III/JG 26 at Ligescourt, while the two remaining 'Richthofen' *gruppen* were at Abbeville-Drucat (II/JG 2), with a detachment at Le Touquet-Etaples under *Hauptmann* Heinz Greisert, and at St Pol-Brias (III/JG 2) under *Oberleutnant* 'Assi' Hahn. The combined establishments of both *Jagdgeschwader* in JAFU 2 totalled only 144 fighters on 28 June, III/JG 26 having been weakened by the absence of its 7.*Staffel* on detachment in the Mediterranean.

Throughout the summer of 1941, the two *Jagdgeschwader* fulfilled their task of defending Northern France more than adequately, although losses, tempered as they were by the advantage of fighting over home ground, were serious enough. The *Kommodore* of JG 2 was killed in combat over Aire on 3 July, and the *Kommandeur* of I/JG 26, *Hauptmann* Rolf Pingel, was captured when he belly-landed his Bf 109F–2 near Dover on 10 July. Losses within JG 26 alone amounted to an average of twelve pilots killed per month over the summer period.

In view of this comparatively high rate of attrition, initial Luftwaffe enthusiasm over the promise of the new Focke-Wulf Fw 190 must have been somewhat dampened when reports percolated through to the units of the problems encountered during the initial service trials with the fledgeling fighter conducted at Paris-Le Bourget. These trials had proved little short of disastrous, the most serious problem having been the soaring cylinder head temperatures in the second row of the BMW 801C–1 radial engine. Both airframe and engine manufacturer had been at loggerheads as a result, and it had been due almost entirely to the energetic action of one man, *Oberleutnant* Otto Behrens, a man of technical background and an active pilot, that over fifty modifications were being implemented in order to render the Fw 190 a viable proposition as a frontline fighter.

Owing to the technical problems plaguing the Fw 190A-1 at Le Bourget, the pilot conversion course proceeded at a snail's pace, but eventually the first Fw 190s released for service were delivered to 6./JG 26 under *Oberleutnant* Walter Schneider at Morseele, a satellite field near Wevelghem. Standard examples of the initial A-1 production model, these mounted an armament of four 7,9-mm Rheinmetall MG 17 machine guns to which had been added a pair of 20-mm MG FF (Oerlikon) cannon, which were installed in the wings immediately outboard of the main undercarriage attachment points. The *Gruppenkommandeur* of II/JG 26, *Hauptmann* Walter Adolph, flew the first Fw 190A-1 taken on charge at the end of July 1941, and 6.*Staffel* quickly relinquished its Bf 109E-7s in favour of the promising new fighter.

By 1 September, the entire II/JG 26 had converted to the Fw 190A-1 and had been declared operational. The pilots were delighted with the handling characteristics

Two Fw 190s in formation over the Adriatic in 1944; nearest the camera is an Fw 190A-3, with an A-4 beyond

of their new mount, and particularly with its incredible rate of roll and diving acceleration. Not so attractive, however, was the choice of weaponry – the MG 17s were good as 'sighting' guns but ineffective in other respects and the low muzzle velocity and rate of fire of the MG FF cannon rendered this weapon far from ideal as fighter armament. Indeed, this combination gave little or no advance on the armament of the Bf 109, and particularly on the F–4 model of the latter, which by now mounted the excellent 20-mm Mauser MG 151.

Although British Intelligence was aware of the existence of the Fw 190, so carefully guarded had been its service introduction that when the new fighter was first encountered, RAF Spitfire pilots assumed that their radial-engined opponent was the aged Curtiss Hawk 75A, examples of which they presumed had been acquired from the *Armée de l'Air*. In their combat reports they referred to a radial-engined fighter which appeared to be fast and manoeuvrable and possessed a performance comparing favourably with the Spitfire VB and the Bf 109F. The first combat sorties had been flown during August, but unfamiliarity with

93

Early morning engine run for Fw 190A–1 *Werk-Nr* 033 of 5./JG 26 at Morseele in November 1941. Number and *Gruppe* bar are black with thin white outline, yellow rudder and overall camouflage of grey. This aircraft was lost at Quehen, near Boulogne, on 22 December 1941 while flying with 6./JG 26 on a transfer flight from Wevelghem to Abbeville-Drucat; the pilot Obfw Kurt Gorbig, was killed

the Focke-Wulf led its pilots to handle it somewhat gingerly, and this had directly resulted in several early combat losses. Within II/JG 26 three pilots of 6.*Staffel* were lost within 10 days – *Oberfeldwebel* Vierling on 19 August, *Feldwebel* Garbe on 21 August and *Leutnant* Heinz Schenk on 29 August – but the most severe blow came on 18 September when the *Gruppenkommandeur*, Walter Adolph, led eight Focke-Wulfs on a convoy escort mission off the Belgian coast.

A shouted warning on the R/T heralded the approach of a strike force of Blenheims escorted by Spitfires, and Adolph led his two *Schwarme* into the attack. A furious mêlée followed just above the sea, during which *Oberfeldwebel* Roth of 4.*Staffel* sent a Blenheim cartwheeling into the water. When the *Schwarme* reformed after the skirmish, however, there was no sign of the *Gruppenkommandeur* and it was initially assumed that he had returned to Morseele. It was only after landing that they learned that Adolph had not returned and was assumed to have crashed. Still uncertain of the operational status of the Focke-Wulf fighter, the official British communiqué relating to the action referred to the destruction of 'a Curtiss Hawk (or Fw 190)'. Three weeks later Adolph's body was washed ashore at Knocke.

Battle returns to *Luftflotte* 3 Headquarters for 25 October 1941 showed that II/JG 26, now under *Hauptmann* Joachim Muncheberg, had fifty-five Fw 190As on strength at Wevelghem, of which thirty-four were combat ready, while III/JG 26 under *Hauptmann* Josef Priller and based at the coastal airfield of Coquelles could muster thirty-eight Fw 190As but only seven of these were fully serviceable, the *Gruppe* remaining operational on the Bf 190F–4 during conversion to the Focke-Wulf.

Good fortune continued to elude the Focke-Wulf-equipped *Gruppen* of JG 26. During a fighter sweep by Spitfires on the evening of 8 November over the area St Omer–Cassel–Hazebrouck the Focke-Wulfs of JG 26 were again in action and during a protracted fight lost three of their number. There was a collision over Ypres and *Unteroffizier* Kern flying Fw 190A-1 *Werk-Nr* 10021 from 4./JG 26 was killed; Fw 190A–1 *Werk-Nr* 10064 was badly shot up and suffered 50 per cent damage in crash-landing at Coquelles, and *Oberleutnant* Theo Lindemann of 6./JG 26 made a forced landing near Dixmuide causing extensive damage to Fw 190A–1 *Werk-Nr* 10052. A little more than a month later, on 22 December, 6./JG 26 was transferring from Wevelghem to Abbeville-Drucat when the *Staffel* became lost in unforecast inclement weather. Visibility was appalling and, circling in the fog over the Artois hills with fuel running low, the *Staffel* quickly lost cohesion, scattered and five pilots lost their lives, including the *Staffelkapitän*, *Oberleutnant* Schneider, who crashed near the Steenvoorde canal in Fw 190A–2 *Werk-Nr* 217.

The first few months of operational service had hardly been auspicious for the fledgeling fighter but the tide of fortune was on the point of turning. Intensive effort had resulted in the teething troubles that had haunted the Fw 190A to that time being largely cured, and the new Fw 190A–2 with the improved BMW 801C–2 had begun to reach JG 26 from the end of November, with *Hauptmann* Priller and *Hauptmann* Muncheberg flying *Werk-Nr* 206 and *Werk-Nr* 209 respectively. Armament had been standardised within the *Geschwader* on two fuselage-mounted MG 17 machine guns and a pair of 20-mm MG 151 cannon in the wing roots. Much was to be made of the hitting power of the early Fw 190 but, in truth, its weight of fire was inferior to that of the standard Spitfire VB by which it was principally opposed, the phenomenal rate of fire of the MG 151 being reduced, as was also that of the MG 17, by synchronisation for firing through the propeller arc. However, the Fw 190A–2 carried more ammunition for its cannon than did the Spitfire VB for its 20-mm Hispano 404 Mk II weapons, the former having approximately 17 seconds of firing time as compared to the 10 seconds of the latter. On both fighters the armourers usually harmonised the guns with a convergence point between 200 and 270 yds (180–245 m).

During the winter of 1941–2, the two *Jagdgeschwader* in the West redeployed to bases that were to be, for some time, their

Early Fw 190As of 7./JG 2 '*Richthofen*' operating in France, circa 1942. In common with most fighters operational during World War II, the Fw 190 operated from grass or dirt fields without difficulty

Focke-Wulf Fw 190A–4/U4s of an unidentified unit operating in France in 1943. The mottle camouflage was unusual for Fw 190s at this period

permanent quarters. The Staff of JG 26, under *Major* Gerd Schöpfel, remained at Andembert, at the western boundary of the rolling hills of Blanc Nez, while a few miles to the east, at Coquelles, was III/JG 26, whose *Kommandeur*, Josef 'Pips' Priller, was by now credited with fifty-eight 'kills' in combat. South of the St Omer forest, at the well-equipped airfield of Arques, I/JG 26 under *Hauptmann* Johannes Seifert was still operating the well-tried Bf 109F–4, with re-equipment to the Fw 190A pending, and II/JG 26, led by *Hauptmann* Jochen Muncheberg, holder of the Oak Leaves and the star of the *'Schlageter' Geschwader* with sixty-two 'kills', was based at Abbeville-Drucat.

From some aspects, when the year 1942 dawned, the Focke-Wulf fighter had still to be thoroughly blooded in combat. Confidence in the Fw 190A had been established, if not without some traumatism, and the western-based *Jagdgruppen* now harboured few doubts as to the ascendancy of the Focke-Wulf Fw 190A over its principal adversary, the Spitfire V, which could be outflown on every count other than turning circle. Perhaps the first real opportunity for the Focke-Wulf pilots to demonstrate their growing *élan* and the mettle of their mount came on 12 February 1942. On that particular morning the weather was foul, with a frontal depression sweeping in from Brittany and Southern Ireland. The cloud base over the Channel and Northern France hung at 1,500 ft (457 m), dropping in some areas to 300 ft (91 m), with flight visibility ranging from 3 miles (4,8 km) down to a few hundred yards in rain squalls. An ideal day for RAF Fighter Command 'Rhubarb' interdiction operations over France.

At 11.00 hours Group Captain F. V.

Another view of Fw 190A–4/U4s in France in 1943. These were reconnaissance-fighters with cameras in the rear fuselage, indicated by fairings just visible in the underside under the canopy

Beamish and Wing Commander R. F. Boyd, both from the Kenley Wing, were returning from an uneventful sortie over the rain-sodden fields of France, flying just south of the Somme Estuary. Crossing the cliffs near Le Treport, they pulled up to fly just clear of the grey cloud base, when, to their amazement, they saw evidence of a large naval convoy. There were three large capital ships with a considerable force of destroyers and fast E-boats providing a screen and all proceeding up coast at high speed. Beamish immediately broke radio silence and informed Kenley Operations of the sighting and hardly more than seconds later the German *Horchdienst* (listening service) informed *Oberst* Galland at his command post near Audembert that the British had sighted the naval force some 7 miles (11 km) off Le Treport.

For the first time that morning the code words 'Open Visor' crackled in the headphones of about forty German fighter pilots as they cruised a few metres above the sea. They promptly adjusted their harnesses and gunsights, and checked their guns. Operation 'Cerberus' or 'Donnerkeil' was no longer a secret! For months the two capital ships *Scharnhorst* and *Gneisenau*, together with the cruiser *Prinz Eugen*, had been bottled up in the French port of Brest; their sojourn had been costly for both adversaries – for the Germans these vessels had been in effect *hors de combat* and for the British their presence had necessitated the maintenance of a powerful monitoring force of warships as well as constant availability of a considerable RAF force for use in the event that the warships were tempted to leave harbour. Unbeknown to British Intelligence, the *Führer* had ordered them out and, after weeks of meticulous planning,

Above, another view of an Fw 190A–4/U4 in France in 1943 (see also pages 96 and 97). Below, Fw 190F–2 fighter-bombers, with tropical filters, of *Gefechtsverband Druschel*

they had finally sailed for Kiel and Wilhelmshaven, following a route of almost suicidal chance – through the Straits of Dover in daylight!

Adolf Galland had been entrusted with the vital task of providing fighter cover from dawn on the chosen day, this occurring when the force was abeam of Cherbourg, until nightfall, which, assuming that all went well, would find the vessels approaching Walcheren on the Dutch coast. The weather had been well chosen and Galland had 252 Fw 190As and Bf 109F–4s at his disposal, all drawn from JG 2, JG 26 and the *Jagdschule Paris,* in addition to a *Gruppe* of long-range Messerschmitt Bf 110s of ZG 26 which was to cover the night-dawn phase of the passage off Cherbourg. The fighter 'umbrella' was planned in such a way as to have a minimum of sixteen aircraft over the naval force at any one time, with an overlap of 10 minutes, so that some thirty-two fighters would be in the neighbourhood for most of the time. Once the code words 'Open Visor' had been given, R/T chatter was authorised and the fighters split into low and high cover around the convoy.

Incredibly, nothing happened for another hour and a half.

The *Kommodore* of JG 26, *Major* Schöpfel, had decided that for this operation he would fly with his erstwhile *Gruppe*, III/JG 26, and after two circuits of Coquelles the Fw 190As of 8. and 9.*Staffeln* swung out over the dunes and low over the Channel to relieve elements of JG 2. By 12.45 hours the naval force was just south of Gris Nez. Shore batteries around Dover were giving a good display of gunnery, with hugh plumes of water marking the fall of shot, when 9./JG 26 saw the first British aircraft in the form of six antiquated Swordfish torpedo-bombers of No 825 Squadron, Fleet Air Arm, covered by the Spitfire VBs of No 72 Squadron from Biggin Hill. The Focke-Wulf pilots had to lower their undercarriages and dump flap in order to stay with the lumbering torpedo-bombers as they came in at 110 mph (177 km/h) just above the waves. The bravery of the Swordfish crews was incredible, but despite the frantic efforts of Squadron Leader C. B. F. Kingcombe's No 72 Squadron to protect them, they succumbed one after another, the

99

An Fw 190A-4 of 1/JG 54 *'Grunherz'* on the Eastern Front in the winter of 1943, operating from a snow-covered airfield

first to go being Lieutenant-Commander Eugene Esmonde, the leader, whose Swordfish had its upper mainplane torn off by a withering hail of cannonfire from an Fw 190 sitting 70 yds behind, shuddering on the stall.

There then followed a free-for-all between the Spitfires, Messerschmitts and Focke-Wulfs which was to continue throughout most of the afternoon in very poor visibility, although the cloud base had lifted to about 5,000 ft (1 524 m) with broken cloud between 1,500 and 2,500 ft (460 and 760 m). Both sides were to make exaggerated claims, owing to the confused nature of the fighting, but at the end of the day the Lufwaffe losses were very light. The *Jagdgeschwader* 2, which had covered the initial and final daylight stages of the passage, had lost seven Bf 109F-4s as a result of crashlandings due to the inclement weather, the aircraft ranging from total write-offs to relatively minor damage and no fatal casualties being sustained. JG 26 was not quite so fortunate as, apart from a Bf 109F-4 from 3.*Staffel*, it lost three Focke-Wulfs from 9.*Staffel* together with their pilots – *Oberfeldwebel* Koslowski in Fw 190A-2 'Yellow 1' *Werk-Nr* 2068, *Oberfeldwebel* Starke in Fw 190A-1 'Yellow 4' *Werk-Nr* 089 and *Unteroffizier* Stavenhagen in Fw 190A-1 'Yellow 12' *Werk-Nr* 042. The following day the British Press released the figures for casualties suffered by the RAF and FAA – twenty bombers, six torpedo-bombers and sixteen fighters. Galland and the *Jagdflieger* had done their work well and the Fw 190A had particularly distinguished itself.

If the British had suffered a bitter blow to morale as a result of Operation 'Cerberus' or 'Donnerkeil', it was not to be reflected by any reduction in offensive operations across by Channel by RAF Fighter Command and No 2 Group of Bomber Command, and as weather improved with March, the round of 'Circus', 'Ramrod' and 'Roadstead' operations recommenced and the Fw 190A was to be encountered with increasing frequency. By mid-March, I/JG 26 had begun to replace its Bf 109F-4s with the Fw 190A-2, completing the conversion of the entire *'Schlageter' Geschwader* to the Focke-Wulf fighter, although *Major* Schöpfel's *Geschwaderstab* retained its Messerschmitts until mid-April, with the fighter-bomber element, 10.(Jabo)/JG 26, following suit in June.

The Luftwaffe Quarter Master General's return for 18 April 1942 listed the *Geschwaderstab* at Audembert with an establishment of five Fw 190A–2s; I/JG 26 at Arques under *Hauptmann* Seifert was recorded as having twenty-nine Focke-Wulfs on strength, II/JG 26 at Abbeville-Drucat under *Hauptmann* Muncheberg had thirty-one Fw 190As and at Wevelghem-Courtrai *Hauptmann* Priller's III/JG 26 had a total of 30 Fw 190As – a relatively small but nonetheless potent force based between Lille and the Somme Estuary under *Jagdfliegerführer* 2. West of the Seine the *'Richthofen' Jagdgeschwader* under *Oberstleutnant* Walter Oesau was still operating the Bf 109F–4 but preparations had reached an advanced stage for the phase-out of this type in favour of the Focke-Wulf and, in fact, II/JG 2 at Beaumont-le-Roger under *Hauptmann* Helmut-Felix Bolz had received the Fw 190A as of 10 May. June found *Hauptmann* 'Assi' Hahn's III/JG 2 at Théville with the Fw 190A, and as of 10 July, I/JG 2, less 1.*Staffel,* was based at Tricqueville under *Oberleutnant* Erich Leie, also with the Focke-Wulf fighter, the *Geschwaderstab* finally taking on the Fw 190A during the course of August. The *Staffelkapitän* of 1./JG 2, which had been detached to Ligescourt to bolster JAFU 2's forces, was *Oberleutnant* Rudolf Pflanz, and this *Staffel* retained its Bf 109F–4s until, in early July, it became the first unit in the West to receive the new BF 109G–1, being renumbered as 11./JG 2 in the process.

From April 1942 onwards, 'Focke-Wulf' was to become something of a watchword for the Allied fighter pilot, conjuring respect and not a little apprehension for the lethality of the German fighter. Against the Bf 109F the pilot of a Spitfire VB knew himself to be pitted against a measured foe likely to afford few surprises, but against the Fw 190A he was aware that he was facing a particularly dangerous adversary – one

An Fw 190A–5 of I/JG 51 *'Molders'* taxies in the slush of the Russian thaw of Spring 1943. This aircraft, 'White 10', is camouflaged with a uniform dark green spray over basic pale blue with yellow fuselage band and rudder base, whereas the example in the background has random white patches sprayed onto the green camouflage

Above, Fw 190A–5/U–3 *Jabos* in the field, with SC 250 bombs loaded. Below, bombing-up a tropicalized Fw 190A–5

whose shortcomings he knew little about but whose positive attributes were likely to be displayed in no uncertain fashion, particularly with a confident and skilful pilot at its controls. RAF Fighter Command losses in March of thirty-two Spitfire VBs shot down and a further twenty-three badly damaged during offensive operations over the continent rose in April to a staggering total of 103 Spitfires and a solitary Hurricane. Admittedly, losses fell during the following month to sixty-one Spitfires and four Hurricanes but attrition was still undeniably high.

The reason for this high attrition was twofold – adequate advanced warning of the mounting of a sortie by the RAF provided by German radar, and the capabilities of the Fw 190A. The German-occupied coastline from Leeuwarden in Holland to Brest on the Brittany coast had been dotted with observer posts and radar installations. The *Freya* and *Würzburg* radars were the main-line defences within the growing framework of the 'Kammhuber Line'. Operating within the 118–130 MHz frequencies, the *Freya* radars were spaced at intervals of about 70 miles (112 km) along the coastline and their range depended on the altitude of the target. For example, the *Freya* at Bellevue – on the high ground behind the cliffs of Blanc Nez and operating on 123·6 MHz – could 'paint' a target flying at 1,000 ft (305 m) at a maximum range of 56 miles (90 km); an aircraft flying above that altitude could be spotted within the arc Eastbourne–Tonbridge–Southend, while one flying above 5,000 ft (1 524 m) could be detected by *Freya* up to a range of 98 miles (158 km). Operating alongside *Freya*, the *Würzburg* – with a dish aerial using the 545–570 MHz band – had a much shorter range but had the advantage of indicating altitude.

Until mid-May 1942, it had been the policy

of RAF Fighter Command and No 2 Group of RAF Bomber Command to rendezvous over a planned point – Newhaven, for example – while orbiting in a shallow climb up to 5,000–7,000 ft (1 525–2 135 m), before finally setting course for the day's selected target. Despite the use of diversionary sweeps and feints, the *Freya* almost invariably gave the *Jagdfliegerführer* adequate warning to scramble the Fw 190s and the numerically fewer Bf 109Gs in good time and to assemble them over St Omer and the Somme Estuary. Even if the Luftwaffe controller erred in his estimation as to which formation constituted the main threat, whichever RAF formation was bounced by the waiting German fighters had to contend with the tactical disadvantage of height and sun. In an effort to elude *Freya* detection, the method of a rendezvous below 500 ft (152 m) – below enemy radar – in complete R/T silence was adopted, the participating aircraft flying at sea level after setting course and then, at a pre-planned time, pulling up in a steep climb as the coast was approached in order to get above 4,500 ft (1 370 m) before crossing the coastal Flak belt.

By use of this tactic an element of surprise was obtained which rendered German reaction late or at a lower altitude, helping to cut the prohibitive losses experienced in April, but this was countered to some degree by the willingness of individual Fw 190 unit *Gruppenkommandeure* and *Staffelkapitäne* to accept combat from lower altitudes and to stay and fight it out (the Messerschmitt Bf 109 pilots tended to display less pugnacity if lacking an altitude advantage and thus unable to employ their preferred dive-and-climb tactics). A typical day during the summer of 1942 was 1 June, a Monday during which the RAF flew over 600 offensive sorties and 53 air-sea rescue patrols over France, Belgium and the Channel. No 11 Group lost ten Spitfire VBs with eight pilots killed or missing, including the Wing-Commander Flying of the Debden Wing, and several more Spitfires staggered back to base with severe battle damage. Three Fw 190s were claimed destroyed but it was subsequently to be revealed that no German fighters had, in fact, been lost!

If not a débâcle for No 11 Group, this particular action served to underline the overall superiority of the Fw 190A to the Spitfire VB and increased the consternation

A bomb-carrying Fw 190 taxying on the Eastern front, where the type became widely deployed after its initial debut in Europe

103

that was widely felt in RAF Fighter Command, bidding fair to affect morale adversely if steps were not taken immediately to redress the balance. Analyses of the operation gave cold comfort to those whose task it was to persuade line pilots that there was no mystique attached to the Focke-Wulf fighter's success, and that it could be countered effectively by the Spitfire VB by application of the correct tactics and exploitation of the new Luftwaffe fighter's weaknesses – a major problem at this time was, of course, that the RAF was unsure as to what the Fw 190A's weaknesses were!

On 1 June, the weather over the Channel had been superb. There were small amounts of cumulus up to 8,000 ft (2 440 m) and above this it was clear with some high cirrus upwards of 25,000 ft (7 620 m). At 10.30 hours the Biggin Hill and the North Weald wings had set off on 'Rodeo 65', sweeping from Hardelot into the Calais area with the aim of goading the yellow-nosed Focke-Wulfs of *Hauptmann* Seifert's I/JG 26 into the air from Arques and thus ensuring that they would be firmly back on

Above, Field Marshal Erwin Rommel inspects the Fw 190A–6 of *Oberstleutnant* Josef Priller, *Kommodore* of JG 26 *'Schlageter'*. Below, Priller climbs down from his aircraft at Lille-Nord after a sortie in the Spring of 1944. Priller claimed 96 combat 'kills' at this time – all achieved in the West. Note the back-pack parachute with rip-chord routed over his shoulder and the absence of a Mae West; the dinghy was strapped beneath the pack and inflated by a CO bottle carried on the left rump

Above and below right, two views of Fw 190F-2s operating with *Gefechtsverband Druschel* on the Eastern front; the example shown below has a tropical filter, necessary for summer operations in Russia. Note in the photo above the absence of wheel well covers on both aicraft

the ground and refuelling two hours later when the planned Circus operation to Bruges followed across the Channel. However, the German controller had kept *Hauptmann* Seifert's *Gruppe* at readiness at Arques; nothing had been seen and the disgruntled RAF pilots had landed back at their bases at 11.30 hours.

Circus 178 had been the first of three similar operations to be flown that day. The order of battle had been impressive – eight Hurricane IIB fighter-bombers from No 174 Squadron each toting a pair of 250-lb (113,4-kg) GP bombs fused for three seconds delay, with a close escort of thirty-six Spitfire VBs from the Hornchurch Wing, escort cover consisting of forty-eight more Spitfires from Biggin Hill, and the target support comprising yet a further forty-eight Spitfires from the Debden Wing. A diversion had been mounted in the Gravelines–Nieuport sector by thirty-six Spitfires from Kenley. At 13.01 hours the Hurricanes had met their escort and escort cover over Eastchurch, and after crossing the Channel at an altitude of 100–200 ft (30–60 m), the Hurricane leader had eased up into a climb in order to cross the Belgian coast west of Blankenberge at 9,500 ft (2 895 m) to the

accompaniment of mild Flak. The Spitfires had been stepped up at various heights from 10,000 to 26,000 ft (3 050 to 7 925 m), with the Hornchurch Wing under Wing Commander R. P. R. Powell staying close to the fighter-bombers.

No reaction had as yet been seen on the part of the Germans. Over Bruges the Hurricanes had peeled off and dived in line astern on their target, the lock gates on the canal, crossing out west of Wenduyne at ground level and suffering no casualties from the barrage of 20-mm and 37-mm Flak. There had been no sign of any German fighters throughout the entire mission, but for the Debden Wing providing target support it had been a different story. The German controller had apparently been over-cautious on this particular morning and it had not

105

Two Fw 190A fighters demonstrate the *Rotte* – the basic combat pair formation, over the Adriatic in 1944. For combat, the formation would open out to a distance of 70–100 yards (74–91.5 m), each aircraft giving the other mutual protection. 'Yellow 8', an Fw 190A-3 in the foreground, is camouflaged in two-tone green with white spinner and yellow rudder and cowl: 'Yellow 9', an Fw 190A-4, is similarly finished

been until 13.20 hours that III/JG 26 had been scrambled from Wevelghem and Morseele, the *Gruppenkommandeur*, 'Pips' Priller, leading the *Gruppe* in Fw 190A-2 *Werk-Nr* 5310. A few minutes later Seifert's I/JG 26 had been climbing in a wide orbit south of Clairmarais forest, his 25 Fw 190As converging on Priller's *Gruppe* as they headed north-east for Ostend. Throttles wide open and climbing steeply, both *Gruppen* had reached 30,000 ft (9 145 m) as they crossed the Belgian coastline between Ostend and Blankenberge, some of the Focke-Wulfs leaving long vapour trails just above the cirrus layer.

The Debden Wing's forty-eight Spitfire VBs had been in battle formation at between 20,000 and 25,000 ft (6 095 and 7 620 m) off Blankenberge when their pilots had first seen the vapour trails approaching fast from the south-west. Twelve of the Focke-Wulfs had come down in a fast feint and the Debden Wing had turned towards them but the German fighters had pulled away. It was during this turn that No 111 Squadron, led by Squadron-Leader P. R. W. Wickham, had begun to fall behind and in order to pick up speed had nosed down slightly. Momentarily the tight defensive formation of the Wing had been lost and No 111 had been immediately engulfed by Fw 190s, the squadron exploding in all directions on the 'break'. Squadron-Leader G. C. Peterson at the head of No 71 'Eagle' Squadron, now flying just above No 111, had committed his unit to the attack and had gone down to assist the beleaguered companion squadron, as, simultaneously, the

Belgian No 350 Squadron that had been flying top cover was 'bounced' by twenty Focke-Wulfs that plummeted down out of the eye of the sun. Spitfires and Fw 190s were fairly evenly matched numerically if not in capability.

Early in the mêlée Wing-Commander J. A. Gordon was heard to call 'Mayday', but both he and his No 2, Sergeant Parrack, had been lost. Sergeants Bryson and Cummings of No 111 Squadron had been shot down in the initial 'bounce'; No 71 had lost Pilot-Officer Teicheira in a winding-match over Ostend; No 350 Squadron had lost Pilot-Officer Laumans, Flight-Sergeant Livyns and Sergeant Hansez; Pilot-Officer Richards had baled out following an attack by a single Fw 190 just off the coast and Sergeant Kopecek, attacked by the same Focke-Wulf, had been badly wounded and had only just succeeded in staggering back to Manston for a crash landing. Many other Spitfires had suffered varying degrees of damage and, despite three Fw 190s claimed, not one of their antagonists had, in fact, been shot down.

Numerous Allied fighter pilots were to submit claims for Fw 190s destroyed in combat on the evidence of white vapour and flames from the engine which gave the impression of a bona fide 'kill'. The Focke-Wulf was structurally very robust, however, and its BMW 801 engine – while then still providing maintenance personnel with some headaches – could absorb a surprising amount of punishment. On one occasion *Leutnant* Victor Hilgendorff had put his Fw 190A down at Audembert after a dogfight with black smoke pouring from the engine. On inspection it had been found that a 20-mm cannon shell strike had blown off two centimetres of one of the cylinders yet the engine had continued to function perfectly. During the testing of the early Fw 190s at Rechlin several pilots had noted that, under certain atmospheric conditions

107

and at certain high power settings, the exhaust of the BMW took the form of a thick white smoke trail which began a short distance from the stubs and hid both wings, giving, as one test pilot put it, 'a decidedly weird feeling'.

Of Debden's casualties, *Hauptmann* Priller had accounted for one Spitfire at 13.45 hours, some 3 miles (5 km) off Blankenberge, this constituting his seventy-third combat 'kill'. Of I/JG 26, which had first 'bounced' No 111 Squadron, the *Gruppenkommandeur*, Johannes Seifert, had shot down a Spitfire 9 miles (15 km) NNW of Ostend, while his *Rottenflieger, Oberfeldwebel* Leibold, accounted for another just over a mile (2 km) North of Zeebrugge. Two Spitfires were brought down in the Ostend area by *Oberleutnant* Schmidt of 3./JG 26, while *Oberfeldwebel* Eichinger of the same *Staffel* had accounted for one over the sea, just NE of Ostend. The seventh of I/JG 26's combat tally had gone to *Feldwebel* Emil Babenz of 2./JG 26. The I *Gruppe* of JG 26 had surely maintained its reputation as one of the most aggressive and successful units on the Channel Front and the danger of the acceptance by Allied fighters of the myth of invincibility that was growing around the Focke-Wulf fighter had been enhanced.

Although the Fw 190A had been on active service on the Channel Front for almost a year, it had only come truly to the fore during the three months preceding June 1942, and it was during this period that it had created most alarm and consternation within RAF Fighter Command. The marriage of a Merlin 61 engine with a two-stage supercharger to the strengthened airframe of the basic Spitfire VB had given birth to the Spitfire IX – later to be held by many to have been the best Spitfire of them all – but the flow of this more potent development of R. J. Mitchell's classic warplane to the squadrons was as yet barely more than a trickle. Meanwhile, Dipl-Ing Tank and his team at Bremen had not been inactive, having evolved the Fw 190A–3 powered by the improved BMW 801D–2 engine and

A heavily-armoured Fw 190A–8 of IV (Sturm)/JG 3 *Udet* taxies to its camouflaged revetment at Dreux after a mission in June 1944. Losses were so heavy among the lumbering '*Panzerbock*' Fw 190s of IV (Sturm)/JG 3, which moved from Salzwedel to Dreux on 7 June 1944, that the unit was decimated within three weeks of starting operations over Normandy

toting no fewer than *four* 20-mm cannon, in addition to two 7,9-mm machine guns, as standard armament – although most *Gruppen* were, in the event, to fly the A–3 from choice with the outboard MG FF weapons removed.

The new subtype was on issue to both JG 2 and JG 26 in France, and to the Staff and the II and IV *Gruppen* of JG 1 based on Dutch airfields. The last-mentioned *Jagdgeschwader* had been allocated the task of the defence of the German homeland by day, and its four *Gruppen* were distributed between Aalborg, Jever, Deelen, Katwijk, Bergen-am-Zee and Woensdrecht. Most sorties were flown against Allied anti-shipping strikes, but II and IV/JG 1 had had several brushes with the Spitfires of Nos 11 and 12 Groups, although their combat prowess was in no way comparable with that of the '*Richthofen*' and '*Schlageter*' units in France.

With the steady proliferation of the Focke-Wulf fighter, the RAF considered it increasingly vital that an example of this outstanding fighter be acquired so that it could be thoroughly 'wrung out' at Farnborough and its weaknesses revealed, if, indeed, it possessed weaknesses. Various schemes were bandied around for acquiring an Fw 190, most such being ruled out as totally impracticable, but in desperation it was finally decided to mount a commando assault on a Channel airfield at which the Focke-Wulf fighter was known to be based, the commando force being accompanied by several technicians and a senior RAF pilot from Farnborough whose task was to be to fly out the prize while the assault force kept down the heads of the German troops guarding the airfield. This somewhat hazardous operation had reached an advanced stage in planning when, fortuitously, III/JG 2 presented the RAF with a brand new example of the latest Fw 190A–3 model.

On the evening of 23 June 1942, the Portreath and Exeter Spitfire wings had returned

An Fw 190A–2 of II/JG 26, in early stages of deployment of the new Focke-Wulf fighter

from a support sweep over Brittany, during which they had had a short and sharp encounter with some Focke-Wulfs over Ile de Batz. After disengaging, they had flown out at sea level but, unbeknown to the Spitfire pilots, they had been trailed by Fw 190As from the *Gruppenstab* of III/JG 2 from Maupertus and *Oberleutnant* Egon Mayer's 7./JG 2 from Morlaix. As they had approached Star Point on the Devon coast, the Spitfire pilots had just begun to relax when 7./JG 2 curved in out of the low sun without any warning. One of the Focke-Wulfs had collided during the head-on pass with the Spitfire of Wing Commander Alois Vasatko who promptly crashed into the sea, the occupant of the Fw 190A–3 (*Werk-Nr* 0330), *Unteroffizier* Willi Reuschling succeeding in extricating himself from his disintegrating fighter and parachuting into the sea about 7 miles (11 km) ESE of Dartmouth. The combat had been brief, both sides being critically low on fuel, but a lone Focke-Wulf proceeded north-east over Exeter. Four Spitfires were hastily scrambled by the Exeter Wing, two crashing on take-off, a third promptly returning to base with R/T trouble and the remaining Spitfire of No 312 Squadron being summarily despatched by the lone German fighter.

Contact with the intruder had then been lost, but shortly afterwards the exciting

109

The arrival at RAF Pembrey of an Fw 190A–3, *Werk* Nr 5313, on 23 June 1942, was a welcome gift to the RAF, which soon had the aircraft in the air for evaluation – carrying the British serial number MP499

news was learned that this particular Fw 190 had landed at Pembrey, near Swansea. The pilot, *Oberleutnant* Arnim Faber, adjutant of III/JG 2, had become disorientated during the dogfight with the Spitfires, had mistaken the Bristol Channel for the English Channel and, being low on fuel, had landed at the first airfield that he saw in the mistaken belief that it was occupied by the Luftwaffe. It was thus, at 20.35 hours on 23 June, that the Allies acquired what was considered to be one of the most valuable prizes of the air war – a brand new Fw 190A–3 (*Werk-Nr* 5313) without a mark on it!

The British took advantage of this bonanza with creditable swiftness. The aircraft was transported to the Royal Aircraft Establishment at Farnborough, dismantled and thoroughly analysed before re-assembly for flight testing, initially by Wing Commander H. J. Wilson, who had to perform the task without the benefit of Pilot's Notes. Several novel features came to light. For example, all services were electrically operated, including the trimming of the tailplane, while the pilot was relieved of the task of controlling mixture, airscrew pitch, boost and rpm by means of an ingenious *Kom-mandgerät* – a 'brain-box' which executed all these functions automatically. Emphasis had obviously been placed on ease of maintenance in the field, with easily operated access panels to the engine, ancillaries and the airframe, and the standard of workmanship was very high.

On 13 July 1942, the Fw 190A–3 was delivered from Farnborough to the Air Fighting Development Unit at Duxford, and here it was put through a comprehensive series of performance tests, and comparative trials were flown against numerous Allied fighter types, such as the Spitfire VB, the Spitfire IX, the Mustang IA and the Typhoon IB. These trials were completely successful, despite some rough running of the BMW 801, but only served to confirm the outstanding combat flying characteristics of this German fighter.

It was noted that, being a standard Luftwaffe fighter on issue to frontline units, the power of its BMW 801 engine had been de-rated, this being general policy at the time with all units operating the Fw 190A – although some *Gruppenkommandeure* overruled their technical officers – in order to prolong the life of the engine. It is perhaps a measure of the performance superiority of the Focke-Wulf fighter over contemporary Allied types that this policy was maintained in 1942.

The cockpit was considered narrow by Allied standards, but the seating position, with the pilot semi-reclining, was considered extremely comfortable and suited to high g manoeuvres if less suitable for long flights. Although excluding an artificial horizon, the instruments were considered to be well positioned and the controls fell easily to the pilot's hand. The sighting view, when the pilot was seated comfortably in the normal attitude, was considered to be about a half-ring (on the gunsight) better than that of the Spitfire, and the view downwards from the centre of the sight graticule to the edge of the reflector plate holder was about 5 degrees. This had not been achieved by elevating the guns and, in consequence, the line of sight, being entirely due to the nose-down attitude of the Fw 190 when in flight. This was of considerable importance. When in a steep turn, most pilots had to increase 'lead' or deflection by pulling up the nose and thus hiding their target for the few vital seconds during which they fired. With the Fw 190, in terms of deflection allowance, the view downwards gave the pilot an opportunity to keep his target in sight for greater crossing speeds and deflection angles than was usual with other fighters. The pilot was well protected from frontal attack by the engine and the 50-mm armour-glass windscreen, and from behind by his bucket seat, which was 8-mm thick with additional 6-mm plates bolted at the rear. His head and shoulders were shielded by a 13-mm back plate. The search view from the cockpit was considered outstanding, there being no need for a rear-view mirror owing to the excellent all-round visibility.

AFDU pilots found taxiing somewhat difficult owing to their lack of familiarity with the self-centering tailwheel, and on the ground the view from the cockpit was considered little short of appalling. Once airborne, however, it was a very different story. One press on the push button near the throttle raised the 15 deg of flaps employed for take-off and a further button raised the undercarriage. A remarkable feature of the aircraft was the lack of retrimming demanded by the various stages of flight. If trim *was* needed, the tailplane provided it quickly and powerfully. Throughout the flight envelope, even at high speeds, aerobatics were found to be very pleasant, with the ailerons and elevators consistently light and effective. The initial rate of climb from fast cruise was high and the angle steep, and when pulling up after a dive the rate of climb was phenomenal.

In the dive – the essential requirement for most disengagements from combat – the initial rate of acceleration was excellent, and AFDU pilots had no difficulty in attaining 580 mph (933 km/h) TAS while passing through 16,000 ft (4 875 m). The controls, although slightly heavier, remained remarkably light and there was little or no trim change at this very high speed. In common with the Daimler-Benz engine of the Bf 109, the direct injection BMW enabled the pilot of the Fw 190 to push his aircraft into a vicious negative g bunt without the motor faltering.

By comparison with the then standard Spitfire VB, the Focke-Wulf displayed an overwhelming superiority in performance. With the Merlin 45 of the Spitfire operating at 3,000 rpm and +21-lb boost and the BMW 801D–2 of the Fw 190 operating at 2,700 rpm and 1.42 *atas* (+20.9-lb) of boost, the German fighter was found to be between 25 and 30 mph (40 and 48 km/h) the faster at all altitudes up to 25,000 ft (7 620 m) and maintained a margin of about 450 ft/min (2,29 m/sec) throughout. When pulling up into a zoom climb from fast level flight, the superiority of the Fw 190's climb was even more readily apparent, and when pulling up from a dive, the Focke-Wulf was found to draw away so rapidly that the Spitfire VB was virtually left standing.

In the dive it was demonstrated that the Fw 190 could leave the Spitfire VB with ease, particularly during the initial stages of acceleration, and in terms of manoeuvrability there was no gainsaying that the German warplane manifested superiority in all respects save the tight turning circle and in this it was inferior to the Spitfire at all altitudes. When the British fighter attempted to 'bounce' the Focke-Wulf while turning, it was found that its opponent's incredible ailerons enabled it to flick into a diving turn in the opposite direction, the Spitfire's pilot being hard put to follow even if prepared for the manoeuvre and standing almost no chance of pulling the correct deflection.

A variety of tactics were tested in order to ascertain the optimum procedure should the Spitfire pilot find himself under attack from a Focke-Wulf. If the Spitfire was cruising at low speed, it was discovered that even if its pilot spotted the Fw 190 while still well out of range, the German fighter overhauled the Spitfire with ease and that the latter, therefore, had no choice but to use its only superior characteristic – turn rate – in order to evade the attack. If, on the other hand, the Spitfire was cruising at *high* speed (ie, maximum continuous rpm), it was recommended that its pilot initiate a fast shallow dive and thus commit his opponent to a stern chase; the Focke-Wulf would, of course, catch up in time but, in the process, would be drawn a considerable distance from its base. It was further recommended as a direct result of the AFDU's findings that Spitfire VBs always cruised at high speed when over enemy territory or when in areas where there was a possibility of encountering the daunting German warplane.

There was, of course, much hope that the more powerful Spitfire IX, which had been evolved as a matter of extreme urgency, would, if not exactly eliminate the margin of ascendancy enjoyed by the Focke-Wulf, at least do much to redress the balance in the fighter-versus-fighter scenario. The Fw 190A-3 was compared in the same manner with a standard Spitfire IX with the Merlin 61 engine at 3,000 rpm with +15-lb boost. Speed runs displayed that the new Spitfire was marginally faster at certain altitudes. For example, while the Focke-Wulf had a 7–8 mph (11–13 km/h) advantage at 2,000 ft (610 m), this advantage had swung in favour of the Spitfire IX at 8,000 ft (2 440 m) where the British fighter was 8 mph (13 km/h) faster, but had diminished to 5 mph (8 km/h) at 15,000 ft (4 570 m), whilst at 18,000 ft (5 485 m) the Fw 190A-3 had recovered the lead by 3 mph (5 km/h). Once at 21,000 ft (6 095 m) the two aircraft achieved parity in level speed, but at 25,000 ft (7 620 m), the Spitfire again displayed an advantage of 5–7 mph (8–11 km/h).

In climbing there was little disparity between the Spitfire IX and the Fw 190A up to 23,000 ft (7 010 m), although the former displayed a slight edge at 22,000 ft (6 705 m), and as the climb rate of the latter fell off with altitude, the Spitfire IX began to come into its own, proving exceptional at 25,000 ft (7 620 m) and above. From high cruise, a pull up into a climb gave the Focke-Wulf the initial advantage, owing to its better acceleration, and this superiority was even more manifest when both aircraft were pulled up into a zoom climb from a dive. This capability on the part of the German fighter was highly important, for the performance differences between the two aircraft were so slight in many respects that the fighter holding the advantage in the climb would also hold the initiative.

In diving, it was ascertained that the Fw 190A could leave the Spitfire IX behind without difficulty, although the superiority was less marked than with the Spitfire VB, and in manoeuvrability the Focke-Wulf still held most of the cards, although unable to compete on the score of tight turns. The Spitfire could not follow in aileron turns and reversals at high speed, and it was concluded that the most disadvantageous altitudes for fighting the Fw 190A were below 3,000 ft (915 m) and in the band between 18,000 and 22,000 ft (5 485 and 6 705 m).

The somewhat colourless, unemotional AFDU reports were perhaps less dramatic than those submitted by the more loquacious combat pilots encountering the Focke-Wulf in battle and surviving to report on their experiences. Almost without exception, the first thing that they noticed was the swiftness and balance of the Fw 190A's aileron control. At 400 mph (644 km/h) plus the aircraft would flick on to its back or make, with ease, incredible aileron turns that would tear the wings off a Bf 109 or severely strain the arm muscles of a Spitfire pilot trying to emulate his German counterpart. The Spitfire pilot would make a 180 deg change of direction in a vertical dive – the Fw 190A would roll through 360 deg. But the Fw 190A was heavily loaded, and as soon as it initiated a steep turn, the telltale white streamers would appear at its wingtips.

The Focke-Wulf trying to 'mix it' in the classic way of steep turning was doomed, for at any speed and even below the German fighter's stalling speed, the Spitfire could out-turn it. This situation brought about a peculiar style of dogfighting, with the Fw 190As endeavouring to keep on the vertical plane by zooms and dives and the Spitfires doing their best to draw them on to the horizontal plane. Frequently, in the heat of battle, the less-experienced of the Focke-Wulf pilots could not resist the temptation to abandon their mounts' combat-style forte and try a horizontal pursuit curve on a Spitfire, with the result that, before they could recover the speed lost in a steep turn, another Spitfire was turning inside them. Conversely, the German pilot that kept zooming up and down was the recipient of only difficult deflection shots of more than 30 degrees. However, the Focke-Wulf was

Manhandling an Fw 190A into its hanger for servicing in France during 1943. Until the advent of the Spitfire IX, developed in haste to improve the British fighter's performance at medium-to-high altitudes, the Fw 190 was the fighter supreme in European skies

extremely vulnerable during the pull-out from a dive, and if its elevators were used coarsely, a violent high-speed flick immediately occurred, followed by a vicious spin. Recovery from a dive had to be progressive, with the pilot taking care not to 'kill' the speed by 'sinking'.

When in trouble, Focke-Wulf pilots displayed a peculiar habit of starting a zoom climb and flicking the wings of their aircraft in feints to left and right. They would then suddenly half-roll with stick right back and fall out of the sky from under a pursuing Spitfire's nose. Even when this 'disappearing trick' had been demonstrated many times and was well known to their antagonists, they still caught out Spitfire pilots with it time and time again. The Fw 190A was a superlative aeroplane by any standards, but the Allies now knew its defects – no fighter ever designed has been without its operational defects.

By August 1942, all *Gruppen* of JG 2 and JG 26 – with the exceptions of their two Bf 109G-equipped component *Staffeln*, 11./JG 2 and 11./JG 26 – were flying what were, by this time, the well-tried Fw 190A–2 and A–3. The bases of JG 26 had remained unchanged, apart from the transfer of the Staff from Audembert to St Omer-Wizernes in June. Combat attrition suffered by the two *Jagdgeschwader* had been very light: from mid-May until the day before the abortive Dieppe raid of 19 August, JG 2 and JG 26 had lost nineteen and fourteen pilots killed in combat or in accidents respectively, while four of the former unit's pilots and two from the latter had been made prisoners of war.

These Fw 190G–8 fighter-bombers of I/SKG 10 are loaded with one 1,100-lb (500-kg) SC 500 bomb and two 66-Imp gal (300-1) drop tanks and are hastily covered with shrubbery to shield them from the prying eyes of low-flying Allied fighters. Spinners are quartered black and red

German reaction on 19 August 1942 to the Dieppe raid – an expensive 'reconnaissance in force' known as Operation 'Jubilee' – was slow, although elements of JG 26 were in action as early as 06.00 hours, when 1./JG 26 scrambled ten Fw 190s (all available serviceable machines) without prior warning from St Omer-Arques, and subsequently eight fighters were put up by each of 2. and 3./JG 26, with all available pilots flying four or five sorties during the course of the morning and early afternoon. By 10.00 hours Dornier Do 217Es from the Dutch-based *Kampfgeschwader* 2 and Junkers Ju 88As from the 1.*Staffel* of *Kampfgeschwader* 77, later joined by the Do 217Es of II/KG 40 and the He 111Hs of III/KG 53, attempted to penetrate the RAF fighter umbrella over the bridgehead and the considerable number of landing craft and support vessels standing off shore. Most of their attacks were broken up by the numerically superior Spitfires, which also engaged the fighter escort of Fw 190s drawn from I and II/JG 2 and all three *Gruppen* of JG 26. These bombing attempts were reinforced by the strenuous efforts of 10.(*Jabo*)/JG 2 and 10.(*Jabo*)/JG 26, the fighter-bomber *Staffeln* newly re-equipped with the Fw 190A–3/U1 and A–4/U1, but the Focke-Wulfs were given little chance to attack the shipping effectively and were soon embroiled in what was probably to be the biggest air battle on the Western Front.

The RAF lost a total of 106 aircraft, including eighty-eight Spitfires, which was a heavy price to pay for the protection that they afforded the invading force, whereas Luftwaffe losses were put officially at forty-eight fighters and bombers, most of the latter being from KG 2. The *Jagdgruppe* suffering the heaviest casualties was *Oberleutnant* Erich Leie's I/JG 2, which, operating from Tricqueville, lost eight Fw 190s – the *Gruppenkommandeur* had been forced to bale out of his brand new Fw 190A–3 (*Werk-Nr* 0326) during a dogfight over Abbeville in the morning and had been wounded, and two other pilots of his *Gruppe* had succeeded in baling out. The *II Gruppe* of JG 2 from Beaumont-le-Roger under *Hauptmann* Helmut Bolz had lost four Fw 190As and three pilots, while *Hauptmann* 'Assi' Hahn's III/JG 2 from Maupertus had sustained only two pilots wounded. The *Experten* of III/JG 2 had had a particularly good day, with the *Staffelkapitän* of 9./JG 2, *Hauptmann* Siegfried Schnell, claiming five Spitfires, *Leutnant* Sepp Wurmheller claiming seven and the *Staffelkapitän* of 7./JG 2, *Oberleutnant* Egon Mayer, notching up his fiftieth 'kill'. The Bf 109G–1s of 11./JG 2 lost one pilot killed.

With JG 26, the story was much the same – a high ratio of successes to losses. Only six pilots had lost their lives, these including the *Staffelkapitän* of 11./JG 26, *Oberleutnant* Johannes Schmidt, whose Bf 109G–1 was shot down during a low-level dogfight off Dieppe. Claims consisted of fourteen enemy fighters destroyed by *Hauptmann* Seifert's I/JG 26, *Feldwebel* Babenz accounting for three of these and two apiece falling to *Oberleutnants* Hemerichen, Zink and Schmidt. *Hauptmann* Meyer's II/JG 26 claimed ten 'kills', and fourteen more were shared between the Staff and III/JG 26.

At 21.21 hours the last Focke-Wulf had landed and the *Geschwader* had flown a total of 377 sorties in thirty-six separate missions. In the air the day had undoubtedly gone to the Fw 190As, but the jubilation was tempered somewhat by the fact that the day had also witnessed a surprise attack on II/JG 26's lair at Abbeville-Drucat by twenty-two B–17E Fortresses of the USAAF – only three minutes had elapsed between the initial warning and the first bombs actually bursting on the field, so that there had been no time to get all the Fw 190s that were being refuelled and rearmed wheeled or taxied to cover, and the B–17s had left a number of the fighters burning.

Despite the gloomy prognostications of failure, particularly from the British, the USAAF day strategic bombing campaign had commenced two days before the Dieppe débâcle, with an attack on the Rouen-Sotteville marshalling yards by eleven B–17s of the 97th Bombardment Group. Initially the campaign was of an experimental nature for both the USAAF *and* the Luftwaffe, the pattern adopted being akin to the 'Circus' operations, with short-penetration targets, such as Rouen, Amiens and Meaulte, and a close escort of RAF Spitfire IXs. Unused to the tight defensive boxes of the Fortresses, the Focke-Wulf *Gruppen* had to develop new intercept tactics, but on 6 September II/JG 26 succeeded in breaking through the accompanying fighter screen and the *Gruppenkommandeur*, Egon Meyer, despatched a B–17F of the 97th over Flesselles, NW of Amiens at 18.55 hours, while *Oberfeldwebel* Roth of 4./JG 26 brought down another belonging to the 92nd nine minutes later just north of Le Treport. The *Gruppe* suffered no losses, although, during the course of the day, *Leutnant* Spinner of II/JG 2 flying Fw 190A–3 *Werk-Nr* 0526 had been killed near Poix.

The following day, while returning from an abortive attack on Rotterdam, the Fortresses were 'bounced' by elements of II and IV/JG 1 over the North Sea. This *Jagdgeschwader*, based about the German Bight sector of NW Germany and in the Netherlands, had freshly converted to the Fw 190A–3 and A–4, and a running battle developed, as a result of which the Fortresses' gunners claimed that, of the Focke-Wulfs that had harried them, twelve had been destroyed, ten more had probably been destroyed and twelve had been damaged. While the two *Gruppen* of JG 1 involved, although admittedly not emulating the success gained on the previous day by II/JG 26 in that they failed to bring down one of the USAAF bombers, had certainly suffered no such disastrous casualties, only *two* Focke-Wulfs having been lost – Fw 190A–4 *Werk-Nr* 5566 flown by *Leutnant* Endrizzi of II/JG 1 and Fw 190A–3 *Werk-Nr* 5436 flown by *Unteroffizier* Platzer of IV/JG 1. Such wildly exaggerated 'kill' claims were to become commonplace as the USAAF daylight bombing offensive developed – a half-dozen gunners in a Fortress box firing simultaneously at one German fighter would each claim it in good faith if it happened to be destroyed.

An outstanding example of this tendency to overclaim was provided on 9 October, when 108 bombers – from the 92nd, 93rd, 97th, 301st and 306th bomb groups – attacked the Fives-Lille steelworks. The gunners of the mix of Fortresses and Liberators participating claimed on their return the destruction of fifty-six German fighters, the probable destruction of twenty-six more and damage to yet a further twenty – more than the total number of fighters put up to oppose the bombers! A revision of the Lille claims finally gave credit for twenty-one destroyed, twenty-one probably destroyed and fifteen damaged. In fact, the *only* German loss on that day was Fw 190A–4 *Werk-Nr* 7043 flown by *Unteroffizier* Viktor Hager of III/JG 26, who went down over the Chemin de Messines. It was during this Lille mission that *Hauptmann* 'Pips' Priller at the head of III/JG 26 succeeded in bringing down a B–17F of the 306th – which was flying its maiden mission – just south of Lille at 10.35 hours, the attack being carried out from astern while flying Fw 190A–4 *Werk-Nr* 2386. That the Fw 190s had wreaked a considerable amount of damage on the intruding force was indicated by the fact that a damaged B–24 Liberator of the 93rd crashlanded in France and one B–17F Fortress of each of the 92nd and 301st came down in the sea; ten Liberators of the 93rd – which, too, was flying its maiden mission with an aircraft type that was receiving its baptism of fire – regained their base with varying degrees of damage, one displaying some 200 holes of various sizes, and no fewer than thirty-six Fortresses were damaged. Although the bombers had encountered some intense and fairly accurate Flak over the target area, the bulk of the damage sustained had been meted out by the Focke-Wulfs.

For the Focke-Wulf *Geschwader*, however, the interception and destruction of the ever more numerous Fortresses and Liberators was already becoming the most harrowing and demanding task with which they had ever been faced. Stern, high-quarter, low-quarter and beam passes were all tried out in order to discover the most effective means of dealing with these heavily armed bombers. Sheer size led to numerous cases of underestimating range, and the long range and high muzzle velocity of the 0·5-in (12,7 mm) defensive weaponry rendered attacks on the bombers extremely hazardous for the Fw 190As.

The 8th Air Force embarked on its first bombing campaign of strategic importance in November 1942, when it turned its attention to the U-boat ports and installations at Brest, St Nazaire, La Pallice and Lorient. Concurrent with the American day bombing offensive, small though it was at first, lay the German reverse at El Alamein in October, followed by the disastrous campaign in the Soviet Union which was to culminate in the Stalingrad débâcle and the loss of the German 6th Army in February 1943. Previously the Luftwaffe had been able to engage its principal adversaries one by one, but it was now faced with the impossible task of mustering adequate forces and reserves to counter the Anglo-American day and night bombing offensive in the West, the Soviet onslaught in the East and Allied success in North Africa.

At first the 8th Air Force attacks occasioned little embarrassment to *Luftflotte* 3 in France, as its fighter units were capable of dealing with the threat, but in November the *Jagdgruppen* were weakened by the transfer of II/JG 2 *'Richthofen'* with its Fw 190As and two Bf 109G-equipped *Staffeln*, 11./JG 2 and 11./JG 26, to Sicily and then to Tunisia. The threat of an Allied invasion in Southern France led to the trans-

A Focke-Wulf Fw 190G–8 of I/SKG 10 taxies in at Dreux after a mission over Normandy in June 1944

fer of *Hauptmann* Leie's I/JG 2 from Tricqueville to Marseille-Marignane and 10.(Jabo)/JG 2 under *Oberleutnant* Schröter from St André to Istres on 9 November 1942. These units were to remain in the south until the beginning of January 1943.

This pattern of sudden transfer at short notice to fill a gap in the patchwork was to become increasingly familiar for the *Jagdgruppen*. Focke-Wulf units facing the RAF and the 8th Air Force and based in Northern France now consisted only of JG 26 and III/JG 2 under *Hauptmann* Egon Mayer, who had moved from Poix to Vannes-Meuçon in early November to protect the U-boat bases. Mayer's *Gruppe* achieved a notable success on 23 November during an attack by B-17 Fortresses on St Naizaire. Heavy cloud and generally inclement weather had caused many of the intruding bombers from the 91st and 306th bomb groups to turn back to their bases, and only nine B-17s reached the target. It was these that III/JG 26 bounced, Mayer's Fw 190As using frontal attacks to take advantage of the bomber's vulnerability from that quarter. Two of the 91st's B-17s received fatal damage in the first head-on pass and two others were seriously damaged, one of these later crashing near Leavesden. A fourth B-17 was brought down near the target. One of the surviving bombers that succeeded in regaining its base claimed the destruction of seven fighters but, in fact, the only German casualty of the encounter was Fw 190A-4 *Werk-Nr* 7061 flown by *Unteroffizier* Angele of 7./JG 2.

By 31 December 1942, the equipment of the *Jagdgruppen* scattered throughout France, Germany, Norway, the Mediterranean and the Soviet Union comprised ninety Bf 109F-4s, 570 Bf 109G-1s, and G-4s, and 580 Fw 190A-2s, A-3s and A-4s. In terms of capability these aircraft were well matched with the contemporary Allied fighters by which they were opposed, although the interception of USAAF bomber formations had revealed inadequate firepower, this being particularly so in the case of the Bf 109G, with its pair of 7,92-

As the Focke-Wulf Fw 190 became more widely deployed, numerous examples were captured in the field. This example, captured at Gerbini, Sicily, in 1943, is seen here being flown by a pilot of the USAAF's No 79 Fighter Group

mm MG 17 machine guns and single 20-mm MG 151 cannon. The demand for the Fw 190 had, by this time, grown to such an extent that several *Jagdgruppen* in the West were forced to relinquish this favoured mount for the Bf 109G-4 and many units operated a mix of the two types. While this increased the technical load for these units, in terms of operational efficiency the excellent high-altitude qualities of the Bf 109G compensated for the Fw 190A's relatively poor performance above 25,000 ft (7 620 m), the former usually flying as top cover.

The *'Richthofen' Jagdgeschwader* (JG 2) under *Oberstleutnant* Walter Oesau began re-equipping with the Bf 109G from February 1943, while III/JG 26 under *Hauptmann* Geisshardt converted wholly to Bf 109G-3 and G-4 fighters at the beginning of the year. The vital necessity of breaking up the heavy day bomber formations and destroying the B-17s and B-24s piecemeal was reflected in the growing weight of firepower of the two German day fighter types that were now pitted against the 8th Air Force.

The Fw 190A-6 reached the *Jagdgruppen* in June 1943, and was fitted with a battery of four 20-mm MG 151 cannon plus the two MG 17 machine guns as standard armament, and this was further increased in the A-7 and A-8 subtypes by the substitution of twin 13-mm MG 131s for the smaller-calibre MG 17s. A pair of MG 131s was also mounted by the Bf 109G-5 and subsequent versions of the Messerschmitt fighter, while the G-6 made provision for the installation of a fuselage-mounted 30-mm MK 108 cannon. Both the Focke-Wulf and the Messerschmitt were fitted with a plethora of *Rüstsätze* (see earlier chapters) which included additional batteries of 20-mm and 30-mm cannon and Wfr Gr 21 mortars – the 21-cm mortar shells being intended to break up the bomber formations so that waves of interceptors following the mortar-equipped fighters could get at individual bombers – but as weight spiralled upwards, performance suffered. The combat weight of the Fw 190 increased from the 7,716 lb (3 500 kg)

Another captured Fw 190A at Gerbini, Sicily, in this case with tropical filters. The wide yellow band round the fuselage was applied by the US forces to indicate a captured enemy aircraft

of the A-2 model of early 1942 in leaps and bounds and was to achieve 9,422 lb (4 274 kg) late in 1943 for the standard unmodified A-8 subtype without any commensurate increase in power, and the Fw 190A's dazzling performance, although still good, had vanished for ever. The introduction of the Jumo-engined D-series, the *Langnasen-Dora,* in the late autumn of 1944 was to re-establish equality of performance with such Allied types as the P-51D Mustang and the Tempest, but never afford the measure of superiority that the Focke-Wulf fighter had enjoyed in the heyday of its youth.

Although the Channel Front provided the backdrop for the greatest triumphs of the Focke-Wulf Fw 190 during 1942 and 1943 as a day fighter, this sturdy all-rounder had soon broadened its repertoire into the fighter-bomber, night fighter, dive bomber, ground attack fighter and tactical reconnaissance fighter rôles on all fronts to which the Luftwaffe was committed save the Eastern Mediterranean.

Apart from JG 2 and JG 26 based in France, II/JG 1 based at Katwijk in the Netherlands had received about forty Fw 190As by June 1942, and IV/JG 1 was in process of re-equipping with the Focke-Wulf at Bergen-am-Zee. In the North, under *Luftflotte* 5, IV/JG 5 had an establishment of thirty-seven Fw 190A-1s and A-2s in June, although these had all been transferred to I/JG 5 by September 1942, this *Gruppe* of the *Eismeergeschwader* being stationed at various bases according to operational requirements, with Trondheim, Bodo and Bardufoss being utilised for the protection of naval units, and Kirkenes and Petsamo, in the far north, for offensive sorties against the Russians. In 1943, a *Jabostaffel*, 14./JG 5, was to be in action with Fw 190As over the White Sea, where it was to continue to make its existence felt until April 1944, by which time I/JG 5, meanwhile re-equipped with Bf 109Gs, was to be fighting in defence of Germany, leaving IV/JG 5, with a mix of Fw 190As and Bf 109Gs, in isolation in Norway.

The year 1942 had seen the introduction of the Fw 190A-3 on the Eastern Front, initially with I/JG 51 *Mölders*, which had re-equipped at Jesau in September and had become operational in the following month, and I/JG 54 *Grünherz*, which made its début in the Soviet Union with the Focke-Wulf shortly before the end of the year. The career of the Fw 190A as a fighter-bomber had, as previously recounted, begun in the summer with 10.(*Jabo*)/JG 2 and 10.(*Jabo*)/JG 26, and by September the two *Jabostaffeln* were performing small-scale low-level attacks over southern England, tying up an entirely disproportionate RAF force to counter these intrusions. Now the versatile Focke-Wulf further extended its repertoire, making its appearance in the ground attack fighter rôle with the *Schlachtgruppen*, partly equipping II/SchG 1 before the end of 1942; this unit deployed detachments of Fw 190A-4/U3s under both *Luftwaffenkommando Don* and *Luftflotte* 4. By early 1943, the three remaining *Gruppen* of JG 54, II, III and IV, had converted to the Focke-Wulf, with III *Gruppe* reverting to the Bf 109G-4 on transfer to the West in February 1943 in exchange for I/JG 26, which operated its Fw 190A-4s from Rielbitzi, Dno, Schatalowka and Ossinowka until June. This *Gruppe* was to claim 127 combat 'kills' during this period for the loss of nine pilots.

With steadily escalating combat attrition and the failure of the aircraft industry to increase production tempo of the Focke-Wulf fighter as rapidly as had been planned, the Fw 190A was in short supply even for the *Jagdflieger*, who were making increasingly pressing claims on deliveries for units based in the West, which were bearing much of the brunt of the conflict against the increasingly numerous USAAF day bomber formations, and the Bf 109G became increasingly prominent among the fighters meeting 8th Air Force intrusions as a result of *force majeure* rather than choice.

By Spring 1943, Fw 190As were operating with the *Nahaufklärungsgruppe* 13 in Northern France as tactical reconnaissance aircraft and with 5.(F)/123 for longer-range reconnaissance missions, 4.(F)/123 in the same sector later being similarly equipped. As fighter-bombers, Fw 190A–2s and A–3s had been on issue to both 10.(*Jabo*)/JG 2 and 10.(*Jabo*)/JG 26 operating from Wizernes, Caen-Carpiquet and St André, and these had been joined in February by the Fw 190A–4/U8 fighter-bombers of II *Gruppe* of *Schnellkampfgeschwader* (lit: Fast Bomber Group) 10 based at Rennes, this *Gruppe* having been formed at Deblin in Poland during the previous November. The *Stab* (Staff) and I *Gruppe* of SKG 10 were established at Amiens-Glisy, and a III *Gruppe*, formed to bring this specialised fast bomber unit to full *Geschwader* strength, was hurriedly posted to *Fliegerführer Tunis*, ten of its Focke-Wulfs being available at Bizerta-Sidi Ahmed by 10 January 1943.

After some sporadic anti-shipping strikes in the western channel, SKG 10 commenced nocturnal attacks on England in April, its first mission on the night of the 16th–17th proving something of a fiasco in that three pilots of II/SKG 10 landed at West Malling in error. The *Geschwader* was further strengthened by incorporating the two *Jabostaffeln* of JG 2 and JG 26 as 13. and 14./SKG 10 respectively, together with 10.(*Jabo*)/JG 54 as a IV *Gruppe*. Daylight operations were mounted but proved very costly, and by May SKG 10 had reverted to nocturnal activities, the II and IV *Gruppen* departing for Sicily in July to be heavily engaged during the Allied invasion of the island, leaving only I/SKG 10 in the West, where it was to remain until its final amalgamation with SG 4 in November 1944.

For the *Jagdgruppen* the most pressing task throughout 1943 was that of countering the 8th Air Force day bomber offensive over the Reich. The prerequisite was the expansion of the single-engined fighter force on a major scale, and production had risen to well over 1,000 aircraft per month by June 1943 – mostly Bf 109G–6s and Fw 190A–5s and A–6s – largely as result of the efforts

Sitting 10 yards (9 m) behind an Fw 190A–7 over the North German plain, a Mustang pilot of the 8th USAAF blasts with his six Colt Browning 0·5-in (12·7-mm) machine-guns. By the summer of 1944, young German fighter pilots were being sent to front-line units with as little as 100 hours flying time – an inordinate amount of which was spent on the technicalities of flying, to the detriment of gunnery and tactics. This Fw 190 appears to take no evasive action whatsoever

of the *Generalluftzeugmeister*, Erhard Milch. Between 1 January and 1 July 1943, the strength of the frontline *Jagdgruppen* had risen from 1,250 to 1,800 fighters, but this had been offset to some extent by the extraordinarily high wastage rate of fighters in the Mediterranean theatre, which seriously affected aircraft availability for the defence of the Reich.

Existing fighter units were increased in size or withdrawn from other fronts, and new formations came into existence. Now commanded by *Major* Josef 'Pips' Priller, *Jagdgeschwader* 26 *'Schlageter'* had reformed 11.*Staffel* in December 1942 and a 12.*Staffel* was added in April 1943, with a 10.*Staffel* during the following month. Within Germany April had also seen the formation of a new *Jagdgeschwader*, JG 11, from two *Gruppen* of JG 1, and by June this unit was up to full strength, while I/JG 26 and II/JG 2 had returned to the homeland from the Soviet Union and the Mediterranean respectively. In July 1943, *Oberstleutnant* Wilcke's *Jagdgeschwader* 3 *'Udet'* had been transferred to the Reich in its entirety, being followed shortly by elements of JG 27 and JG 53 from the south, but these units were wholly or partly equipped with the Bf 109G as the Focke-Wulf fighter was still in short supply despite accelerating production tempo. The *Stab* of JG 2, for example, had found itself flying the Messerschmitt from February until the late summer of 1943, after which it switched once more to Fw 190As; I/JG 2 was also forced to switch to the Bf 109G in February, operating a mix of this type and the Fw 190A from June, II/JG 2 being entirely Bf 109G-equipped from the early summer until the end of the year and III/JG 2 operating both Fw 190As and Bf 109Gs until May, when the entire *Gruppe* once more flew Focke-Wulfs.

Frontline fighter strength in Germany and the western occupied territories climbed from 635 in January to 975 in October 1943. Allied with the numerical increase were improvements in airfields, fighter control and aircraft armament which brought about notable victories for the Luftwaffe – such as the Schweinfurt/Regensburg attack of 17 August when sixty out of 376 participating bombers were brought down and the Schweinfurt raid of 14 October when sixty out of 294 bombers were destroyed, the latter bringing to a crescendo attrition of no less than 11 per cent suffered by the US bombing force over a period of four days.

Two months later the arrival of the long-range P–51B Mustang coupled with the range increase brought about by the use of the 91·6 Imp gal (416 litre) belly tank on the P–47D Thunderbolt negated the efforts of the Luftwaffe at a stroke. The orderly rendezvous of the *Jagdflieger* over the Dummer See or Steinhuder Meer could now turn into a rout, the efforts of the Focke-Wulfs and Messerschmitts to intercept the bombers frequently being foiled by the 'little friends' by which they were accompanied.

By February 1944, units engaged in defence of Reich territory wholly or partially equipped with the Fw 190 included the *Stab*, I and II/JG 1, *Stab*, I, III and 10./JG 11, II/JG 54 and various new formations, such as the *Sturmstaffel* 1, the *Stab* and II/JG 300 and elements of JG 301 and 302. Fw 190-equipped units based in Belgium and France under *Luftflotte* 3 at this time included the *Gruppenstab* of (F)/123 together with 4. and 5.(F)/123 and NAGr 13 (less the 3.*Staffel*) for tactical and long-range reconnaissance; the *Stab* of *Oberstleutnant* Egon Mayer's JG 2 flying the Fw 190A–7 with the I and III *Gruppen* partially mounted on this type (II/JG 2 being mounted entirely on the Bf 109G); and *Oberstleutnant Priller's* JG 26 of which all but the III *Gruppe* were flying the Fw 190A–6 and A–7. In addition, I/SKG 10 was soldiering on in the fighter-bomber rôle in which it was soon to be joined by III/SG 4.

Meanwhile, there had been something of a paucity of Focke-Wulf fighters on the

On test in the USA in April 1944, a Focke-Wulf Fw 190G–3 sports standard USAAF markings and the tail serial EB101

Eastern Front. Indeed, by February 1944, only *Jagdgeschwader* 54 *Grünherz* was operating the Fw 190A in the air superiority rôle, and while the *Stab* and I and II *Gruppen* were operating the Focke-Wulf exclusively, the remaining component of *Grünherz* in the Soviet Union, the IV *Gruppe*, was operating it in concert with the Bf 109G. All other Focke-Wulfs in the East were flying with the *Schlachtflieger*, these comprising the *Stab* and I and II *Gruppen* of SG 10 and II *Gruppe* of SG 77, other

123

Seconds after 9th USAAF Marauders released their bombs during a mission over West Germany in December 1944, this Fw 190D-9 flashed below in a stern quarter-attack. Note the broad coloured band on the fuselage denoting an aircraft of one of the *Reichverteidigung* formations – the unit was probably III/JG 54

intended to supplant the Ju 87 in the *Stukagruppen*. In the event, despite the urgency attached to the replacement of the elderly Junkers with the Focke-Wulf, the re-equipment programme saw little progress during the summer of 1943, but by October, when the *Stukagruppen* were redesignated as *Schlachtgruppen,* the 87-equipped *Gruppen* were converting to the Fw 190G at a rate of some two *Gruppen* every six weeks – the Fw 190F had been withdrawn from production in favour of the Fw 190G but was to be reinstated in the spring of 1944 and was subsequently to re-equip the remaining Ju 87 elements of the *Schlachtflieger* on the Eastern Front.

The invasion of Normandy on 6 June 1944 found the day fighter strength of *Jagdkorps* II, the operational fighter command in France, at its lowest ebb. The brunt of the initial fighting in the air had to be borne by JG 2 'Richthofen' under *Major* Kurt Bühligen, the third *Kommodore* in as many months! Only a week before the mounting of the Allied invasion the *Geschwaderstab* of JG 2 and the II *Gruppe* could muster only fifteen serviceable Bf 109Gs at Creil. A few miles to the west, at Cormeilles-en-Vexin, the I and III *Gruppen* could put up only a paltry thirty-five Fw 190A–7s and A–8s between them. This shortage of operational aircraft within the *'Richthofen' Geschwader* was almost totally due to the spate of Allied airfield attacks within the sphere of *Jagdkorps* II operations that had formed part of the run-up to the actual invasion.

units operating a mix of Fw 190s and Ju 87Ds, these being II/SG 2 and 1. and 4./SG 5. The growth in importance of the *Schlachtflieger* and increasing evidence of the unsuitability of the Ju 87 for use as a *Schlachtflugzeug* had resulted in the development of a version of the Focke-Wulf fighter optimised for the close support rôle under the designation Fw 190F, while an extended-range fighter-bomber version, or *Jabo-Rei*, had been evolved as the Fw 190G, this being

The establishment and serviceability of the fighters of *Jagdgeschwader* 26 – the only other Luftwaffe fighter unit based in France – remained strong, but on the actual day of the invasion Priller's *Jagdgruppen* were well scattered, with I *Gruppe* at Rheims, II *Gruppe* on leave at Mont-de-Marsan, near the Pyrenees, and III *Gruppe* at Nancy-Essey in Alsace. Only *Oberstleutnant* Priller and his *Rottenflieger, Unteroffizier* Heinz

Wodarczyk, remained in the north with the two Fw 190A–8s of the *Geschwaderstab* at Lille-Nord.

The invasion resulted in immediate redeployment of part of the already overstretched fighter force, and reinforcements arrived quickly in the combat area, with 998 day fighters being included in the total of 1,105 combat aircraft transferred to Northern France between 6 and 7 June. It suffices to say that this redeployment was undertaken at the dire cost of virtually denuding other equally vital battlefronts of fighters. By 10 June the fighter force within *Jagdkorps* II comprised I, II and III/JG 1 based at Le Mans, Flers and Beauvais-Tillé; I and II/JG 11 plus 10./JG 11 based at Rennes-St Jacques and Beauvais-Tillé; II, III and IV(*Sturm*)/JG 3 at Evreux-Fauville, St André and Dreux; *Stab*, I, III and IV/JG 27 based at Romilly-sur-Seine, Rheims-Champagne and Champfleury; III/JG 54 at Chartres; and the autonomous JGr 200 stationed further south, at Avignon. This substantial force served to bolster JG 2 and JG 26, which, meanwhile, had been brought up to full strength.

Aerial warfare over Normandy was intense, with 10,061 sorties being flown by the fighters and fighter-bombers of the Luftwaffe over the Invasion Front between 6 June and 1 July 1944. The *Jagdgruppen* claimed 414 'kills' by day with a further 219 by night, while Flak, by now feared by the Allies far more than German fighters, allegedly accounted for a further 672 Allied aircraft. On the debit side, *Jagdkorps II* lost 485 single-engined fighters and to this total were to be added 224 bombers and a further 137 aircraft destroyed on the ground by bombing or strafing.

Few of the *Jagdgruppen* long survived the holocaust above Normandy as organised tactical units, and within two weeks of D–Day the *Jagdflieger* were being forced to restrict operations to a minimum owing to fuel shortages. But despite the débâcle in Norman skies and contrary to the contemporary Press opinion, the Luftwaffe was far from being a spent force. Expansion of day fighter production had continued to gather momentum under the auspices of the *Jägerstab* (Fighter Staff) – established within the *Reichskriegsministerium* under the leadership of Party Leader Otto Saur – and from the beginning of September to mid-November 1944 the establishments of the *Jagdgruppen* were to rise by a staggering 70 per cent, from approximately 1,900 to upwards of 3,300 day fighters! To cater for this massive influx of fighters, units were enlarged by the addition of a fourth *Gruppe* and the establishments of the *Staffeln* were raised by 50 per cent. The nocturnal *Wilde Sau* units – JG 300, JG 301 and JG 302 – had been turned over to day fighting in June, and after the Allied invasion of Normandy at least six new fighter units were formed. Aircraft attrition in the Fw 190-equipped units had soared, but as by July 1944 the monthly acceptances of the Focke-Wulf fighter handsomely exceeded 1,000 aircraft, this was less of a problem than the shortages of fuel and the losses in aircrew, which had risen to something of the order of 20 per cent monthly.

By the end of October all the *Jagdgruppen* which had met with such total reverses over Normandy had been restored, re-equipped and returned to combat. Furthermore, the seesaw battle for parity at least and ascendancy when possible in fighter performance that had been fought throughout the war by the opposing design teams had now taken another turn with the service introduction of the Bf 109K and the Fw 190D which, forming the new line in improved piston-engined fighters, raised the *Jagdflieger* once more to at least equality of combat performance with the most recently committed Allied types. The *Langnasen-Dora* or *Dora–9*, as the Fw 190D–9 had been dubbed with its arrival in the ranks of the *Jagdgruppen*, was a highly effective fighter of which deliveries began from Focke-Wulf's Cottbus assembly lines in August 1944. It was capable of outclimbing and outdiving its BMW 801-powered predecessors with ease, and the pilots of III/JG 54, the first *Gruppe* to convert to the new Jumo 213-powered Focke-Wulf (and achieve operational status during October), were convinced that the *Langnasen-Dora* was more than a match for the much-vaunted P–51D Mustang.

On 10 December, the *Jagdgruppen* operating the Fw 190 in Germany consisted of the whole of JG 1 *'Oesau'* with A–8s at Greifswald, Tütow and Anklam; the *Stab* and I and II/JG 6 with A–8s at Perleberg and Hagenow; the *Stab* and I and III/JG 11 at Celle and Wunstorf with A–8s; III/JG 54 at Verrelbusch with D–9s; the *Stab* and II/JG 300 at Jüterbog and Lobnitz with A–8s and the similarly equipped three *Gruppen* of JG 301 at Sachau, Stendal and Salzwedel. Focke-Wulf-equipped units under *Luftwaffen Kommando West* comprised the *Stab* (with A–8s) and I and III/JG 2 with D–9s; the *Stab* and IV(*Sturm*)/JG 3 with A–8s; the *Stab* and II(*Sturm*)/JG 4 with A–8s; the *Stab* and I and II/JG 6 with A–8s; the *Stab* and I and II/JG 26 with D–9s; and *Einsatzstaffel*/JG 104 also flying D–9s. In the ground-attack rôle the Fw 190F was being operated by the three *Gruppen* of SG 4 by day and with NSGr 20 by night.

Attrition was high in the desperate air battles fought over the shrinking Third Reich as 1944 drew to a close, despite the fact that flying was inhibited by fuel shortages. Typical was the combat attrition suffered by I and II/JG 26, newly converted to the Fw 190D–9. On 23 December *Leutnant* H. Wirth of II *Gruppe* was forced down near Münster by RAF fighters and

Above and below, an Fw 190D–9 photographed at Rhein-Main airfield in 1945, apparently after making a belly landing but also possibly showing the effects of battle damage. The aircraft was serving with JG 26, with *Reichverteidigung* bands round the rear fuselage

This Fw 190D-9, 'Black 12' of 10./JG 54, was downed by a partridge that caused a hole several inches in diameter in the coolant radiators. The pilot, *Leutnant* Theo Nibel, crash-landed north-west of Brussels and was taken prisoner by Belgian police

on the following day the II *Gruppe* lost four D-9s, three of these in combat with Spitfires and Lightnings over Malmèdy and Rheine, and the other, flown by *Leutnant* S. Benz of 6.*Staffel,* being shot down in error by the Germans themselves. On 25 December, one D-9 flown by *Feldwebel* Hoppe of II *Gruppe* was shot down over Dortmund with the loss of the pilot and another, flown by *Feldwebel* Meindl, was destroyed over Furstenau, the pilot baling out but his parachute failing to open. On the next day, 26 December, the I *Gruppe* lost at least five D-9s over Belgium with the loss of all pilots. Thus, within four days, the two *Gruppen* of JG 26 had lost at least a dozen aircraft in combat and only one pilot had survived. The other Focke-Wulf-equipped *Jagdgruppen* were suffering similar attrition. For example, on 27 December the 10. and 12.*Staffeln* of JG 54, led by *Oberleutnant* Dortemann and *Leutnant* Crump respectively, engaged Tempests of No 486 Squadron, RAF, and P-47D Thunderbolts of the USAAF in the vicinity of Münster and over Telgte, four 10.*Staffel* Fw 190D-9s and one 12.*Staffel* Fw 190D-9 being lost with three pilots killed and two wounded. Three days later the similarly equipped III/JG 54 met its Waterloo in combat with Spitfires of Nos 331, 401 and 411 squadrons and Typhoons of Nos 168 and 439 squadrons, losing twenty of its D-9s with seven pilots killed, including the *Kommandeur, Hauptmann* R. Weiss, a *Ritterkreuzträger* with 121 'kills' to his credit. Most of the pilots killed belonged to 9.*Staffel* which, at the end of the day, was left with one Fw 190D-9.

As was to be expected, the Focke-Wulf fighter played a prominent rôle in Operation *Bodenplatte*, the last concerted offensive action to be mounted by the Luftwaffe; this was an ambitious and ill-fated attempt to attack some sixteen Allied tactical airfields in the Low Countries and north-eastern France simultaneously on New Year's Day 1945. The units involved, which were flying either the Fw 190A-8 or Fw 190D-9 and provided almost two-thirds of the fighters participating in *Bodenplatte*, were I and II/JG 1; I, II and III/JG 2; III and IV/JG 3; I, II and IV/JG 4; I, II and III/JG 11; I and II/JG 26; and III and IV/JG 54. There were, in addition, the Fw 190F-8s of SG 4. The operation, intended as a body blow to Allied air power on the Continent,

An Fw 190D-9 at dispersal during the winter of 1944-45. The D-9 was one of the principal types used in the *Bodenplatte* operation – the last major offensive mounted by the *Luftwaffe* in World War II

was, at best, a forlorn gamble, and proved, in fact, a body blow to the *Jagdflieger* – more than 200 German fighter pilots failed to return from *Bodenplatte*, including a number of highly experienced, irreplaceable commanders. It was a blow from which the fighter arm of the Luftwaffe was never to recover.

Despite the monumental efforts that were still being made to keep the *Jagdgruppen* at a technical and numerical parity with their opponents, time was running out for the Third Reich and neither the quality of Kurt Tank's fighter nor the valour of its pilots could now affect the issue. Added to the overwhelming commitments on all fields by the *Jagdflieger* were the curtailing of vital fuel supplies and the lowering of the standards of the average German fighter pilot owing to the inadequate training dictated by the exigencies of the times. Combat attrition was becoming increasingly insupportable – losses in Fw 190D-9s on one typical day, 14 January 1945, sustained by *Jagdgruppen* operating in a limited area and totalling twenty-eight aircraft represented some 15 per cent of all fighters of this type on firstline strength* – yet, although Soviet forces had by now overrun some 30 per cent of all Fw 190 production facilities, the conversion of units to the *Langnasen-Dora* continued into the last weeks of the conflict. A substantial proportion of the newly delivered fighters, however, were to sit out their brief service careers with empty tanks.

Those for which fuel could be found fought on, contesting with the immeasurably superior numbers of Allied fighters the steadily diminishing German airspace. But whether in the hands of an experienced *Major* who had first seen combat six Springs earlier – now sitting behind a frantically jinking 2nd Tactical Air Force Tempest over the Reichswald in his *Langnasen-Dora*, one eye glued to the fuel gauge – or with a 19-year old *Unteroffizier* at the controls, training just completed and trying not to shut his eyes as his war-weary Fw 190A-8 'Sturmbock' closed with a B-17 at upwards of 600 mph (965 km/h), Kurt Tank's Butcher-Bird acquitted itself well to the bitter end.

---

* I/JG 2 lost three D-9s at Altenstadt, III/JG 2 lost two at Lich and Hagenau, I/JG 26 lost three in the vicinity of Cologne and Lengerich, II/JG 26 lost nine in the same general area, the *Stab of* JG 301 lost one and II/JG 301 is believed to have lost at many as ten in the area Barenthin-Kyritz-Gumtow.

Technical Description

Perhaps the most painstaking technical analysis of any enemy fighter made in wartime was that conducted by the Royal Aircraft Establishment at Farnborough on the Focke-Wulf Fw 190A–3, *Werk-Nr* 5313, during July and August 1942. This aircraft had been captured intact on 23 June 1942 at RAF Pembrey, in South Wales, where it had been landed, apparently in error, by *Oberleutnant* Arnim Faber, adjutaut of III/JG 2. *Werk-Nr* 5313 was a standard A–3 subtype with the new BMW 801D–2 engine, included the full armament of four 20-mm cannon and two 7.92-mm machine guns and had only recently been issued to the *Gruppenstab* III/JG 2 at Maupertus airfield, near Cherbourg. The performance figures, engine output and weights of this frontline Fw 190A–3, therefore, give an accurate picture of the various parameters pertaining to this type which was on issue to the *Jagdgeschwader* from late May 1942 onwards.

It is significant to note that in 1942 the performance of the Fw 190 was considered so superior to current Allied types that orders were given for all frontline Fw 190s in France to fly on combat operations with their BMW 801 engines derated in order to conserve their life. Maximum power was reduced from 1.42 atas (20.9 lb/sq in) boost, which offered some 1,770 hp at rated altitude down to 1.35 atas (19.84 lb/sq in) giving only 1,595 hp at 2,500 ft (760 m) and 1,455 hp at 18,000 ft (5 500 m). As the Allied ability to combat the Fw 190 grew, however,

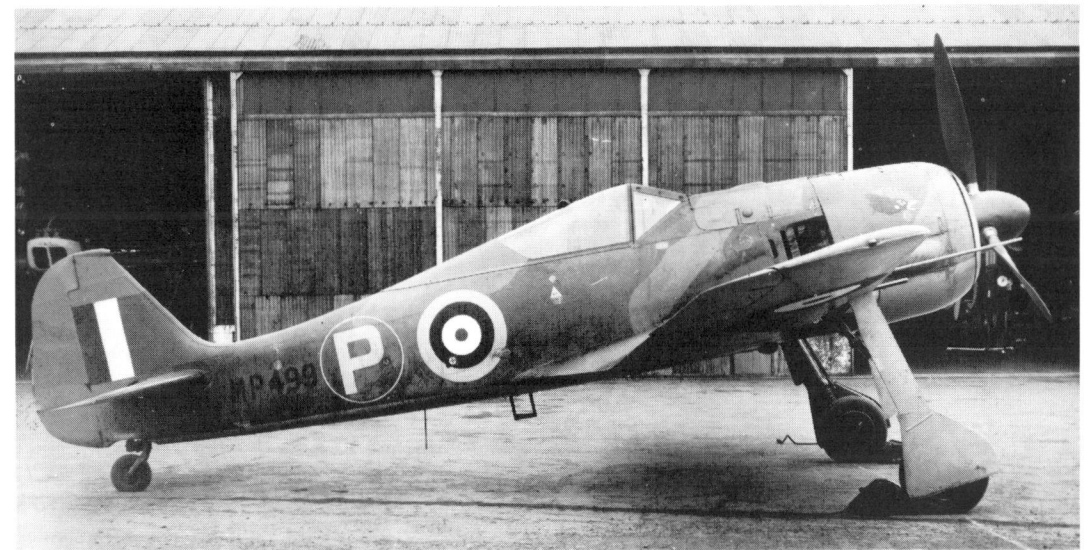

This Focke-Wulf Fw 190A–3, *Werk-Nr* 5313, which was landed in error by *Oberleutnant* Arnim Faber of III/JG 2 in Wales during 1942, was the subject of a full technical assessment at the RAE. The report is summarised in this chapter

this limitation was naturally lifted and the fighter's full performance was used. The description that follows summarises the RAE report on this particular Fw 190.

The Airframe
Wings

The whole wing was made up of one integral unit, with only the moulded wing-tips being detachable: any battle damage sustained by the wing necessitated a complete wing change rather than wasting time on complicated field repairs. The general structure of the Fw 190 wing consisted of a single section main spar, two rear spars and numerous ribs and stringers. The integral centre section of the tapering main spar was a heavy triple-web I-beam extending outboard of the 20-mm MG FF cannon and bore the load of the two side and bottom engine mountings, the main landing gear and the four wing-mounted cannon. At its centre the beam was $16\frac{3}{4}$ in (42,5 cm) deep, reinforced by vertical channel-section members, and it carried the bottom engine mount structure and one of the three fuselage attachment points. Outboard of the side engine mounts the spar was tapered forward 14 deg, the the angle being maintained outboard to the main undercarriage fittings, from which point it paralleled the centre section. At the point of egress from the fuselage the top of the I-beam was bolted to the fuselage firewall (bulkhead No 1). Outboard of the MG-FF gun ports the spar reduced to a single-web I-beam fitted with lightening holes. Attached by screws to the leading edge of the main spar, a separate sub-assembly ran from engine cowl to undercarriage and was made up of stamped flanged ribs with cut-outs for the oleo leg; this portion also contained the heater muff for the barrel of each inboard Mauser MG 151/20 and the undercarriage up-locks. The leading edge, from

Wing Aerodyamic Data	
Area, gross	203 sq ft (18,85 m²)
Area, net	177 sq ft (16,44 m²)
Span	34 ft 6 in (10,52 m)
Mean chord	5.88 ft (1,79 m)
Aspect ratio	5.87
Dihedral	5 deg constant
Sweepback ($\frac{1}{4}$ chord line)	5.5 deg
Chord root	7.50 ft (2,28 m)
Chord tip	4.05 ft (1,24 m)
Flaps	
Type	Split
Max deflection angle	60 deg
Flap area	18.7 sq ft (1,74 m²)
Flap chord/local chord	0.16
inner	0.21
Outer	7 ft 10 in (2,40 m)
Flap span	0.45
Flap span/Span	0.09
Central cut-away/Span	Frise
Ailerons	
Type	20.1 ft (6,13 m)
Aileron area/S	20 sq ft (1,86 m²)
Aileron chord/local chord	0.295
Aileron span/Span	0.43
Percentage balance	28.8
Aileron angles	±17 deg
Stick gearing deg/in	3.2 deg

The Fw 190A-3 on test in British markings as MP499. The German marking has been crudely painted out on the side of the fuselage, with the British roundel painted farther forward. The cockerel insignia on the nose was of German origin

Another view of MP499 while on test from Farnborough in August 1942. In this view, the cockerel insignia has been overpainted

the undercarriage to wing-tip, was made up of formed aluminium sheet reinforced by D-type nose ribs. The rear spars were conventional tapering I-beams which extended from top and bottom fuselage attachment points to the wing-tips, and carried both the flaps and the ailerons; some 32 in (81 cm) out from the fuselage the double web reduced to single web construction.

Five wing ribs, besides those at the tips, were attached to both the top and bottom skins of the wing surface and were placed just outboard of the MG 151/20 cannon bay, on either side of the undercarriage fitting and around the outer MG FF cannon bay; these were conventional stamped, flanged ribs. The top skin surface carried three flanged contour ribs between the 'solid' ribs and a further six ribs between the outer flap hinge and the top rib. All these ribs had cut-outs for Z-shaped spanwise stringers, of which there were nine outboard of the flap and eleven between the 'solid' ribs. Of the bottom skin surface, one contour rib was located at the fuselage attachment fittings, one between the cannon and undercarriage ports, and five between the outer 'solid' rib and the tips; once again, cut-outs were provided for eight spanwise Z-section stringers. Both top and bottom wing panels were covered by flush-riveted Dural, with the sideways-hinged access panels for the MG 151/20 cannon on the top surfaces and the forward-hinged panels for the MG FF cannon on the lower surface, in addition to ports for the flap motors and aileron linkages.

The flaps were built around a metal monospar with top and bottom sections riveted together at the leading edge and the whole covered by doped fabric. They were of the split type, each 7 ft 10 in (2,69 m) in length, and worked by three electrically operated push-buttons on the left cockpit console: a red button raised the flaps, a yellow button selected 10 deg of flap for take-off and a green button selected 60 deg of flap for landing; if the buttons were not pressed fully home and locked, but just held down, any intermediate flap position could be maintained. In addition to flap indicator lights in the cockpit, a small aperture in the top skin panel of the wing gave a dial reading of between 0 and 60 deg. The flaps were electrically driven through a nut to a screw jack attached to the 24-volt motor bolted to the front of the rear spar. The two motors were interconnected through a relay control

Tail unit Aerodynamic Data	
Gross tailpiece area	31.6 sq ft (2,94 m²)
Elevator area	12 sq ft (1,11 m²)
Elevator angles	
Down	26 deg
Up	31 deg
Type of balance	Shielded horn
Percentage balance	9.1
Stick gear	4.1 deg/in
Tail volume coefficient	0.45
Gross fin and rudder area	24.3 sq ft (2,26 m²)
Rudder area	7.65 sq ft (0,71 m²)
Rudder angles	±16 deg
Type of balance	Shielded horn
Percentage balance	4.6
Pedal gearing	6.0 deg/inch

box in order to synchronise the movement of the flaps to within 3 deg.

Fabric-covered metal Frise-type ailerons were as light in weight as they were reported to be on the controls and were built around a channel monospar to which were riveted upper and lower metal leading-edge skins; aft of the spar, there were ten conventional ribs with rounded gussets, and ten stamped, flanged intercostals with the whole aileron covered by fabric and hinged at ribs No 1, 5 and 9.

Fuselage
The fuselage was constructed in two main components, the forward section extending from the light sheet-steel and alloy firewall (bulkhead No 1) to bulkhead No 8, aft of the pilot's seat. The forward fuselage was a double-deck structure separated by a stressed alloy floor, with the upper section providing a forward shelf for the MG 17 machine gun mountings, cockpit space for seat, controls and instruments and further space, aft of the seat, for the FuG 7 R/T radio equipment and 24-volt 10-Amp/hr accumulator. The lower compartment held the main spar with its attachment points, ammunition magazine for the wing-mounted MG 151/20 cannon and two fuel tanks – the forward tank holding 51 Imp gal (232 litres) and the aft 64 Imp gal (292 litres) of C3 (100 Octane) petrol; the tanks were of the self-sealing type suspended from the contour rib-bulkheads by heavy web straps and were replenished through two filler caps on the starboard side of the fuselage.

Aft of bulkheads No 8 to No 14, which held the tail unit, the fuselage contained the oxygen bottles (spherical type), FuG 7 radio generator, FuG 25 IFF equipment, bomb-fusing gear, provision for a camera and master compass sensing unit.

Four fuselage longerons ran aft from bulkhead No 1 to bulkhead No 8, where they were spliced to lighter ones in the aft section. The top longerons were $1\frac{3}{4}$ in (55,5 mm) wide U-sections which served as tracks in which the cockpit canopy travelled. One hat-shaped stringer on each side, riveted $10\frac{1}{2}$ in (26,7 cm) below the top longeron, constituted the only horizontal stiffeners in the top fuselage, while the cockpit floor was joined to the bottom longerons, separating the cockpit from the fuel bay. The bulkheads in the upper fuselage section were not of the same construction and were unevenly spaced. Bulkhead No 5, directly under the rear end of the fixed windshield and above the rear spar fitting, extended above the floor only to the stringer and was braced by a $\frac{3}{4}$ in (19 mm) tubular section riveted to the bulkhead and the cockpit floor. No 6 Bulkhead was an A-frame which supported the pilot's seat and armour plate.

The lower fuselage or fuel bay had six bulkheads, with No 4 being the tie-through member for the rear spar fittings and sep-

In the three years after the arrival of Arnim Faber's Fw 190 at the RAE, several other variants of the Focke-Wulf fighter were tested at Farnborough, among them this rare example of the two-seat Fw 190A–8/U1

Another rare Fw 190 that was seen at the RAF, Farnborough, during 1945 – although not test-flown there – was this *Mistel* S3A composite intended as a trainer for pilots destined to fly the *Mistel* on operations, in which the lower component comprised a surplus bomber airframe packed with explosive

arating the two fuel tanks. One large belly skin panel was screwed to the underside of the fuselage to give easy access to the fuel tanks. Of the after portion of the fuselage, which continued backwards from bulkhead No 8, the construction was a semi-monocoque structure built up of bulkheads and several Z- and U-section stringers; No 12 contained a fabric panel to prevent dust seepage to the accessories while No 13 contained a cross-tube for lifting the fuselage.

Attached to the front of the firewall were the ammunition magazines for the Rheinmetall MG 17 machine guns, each holding 1,000 rounds; vertical chutes routed the belted ammunition to the weapons, which were mounted on the forward fuselage shelf alongside a compressed air bottle and associated fuse-boxes. A large fairing, extending from the windshield to the engine cowl, covered the weapons and was constructed of heavy, waffle-bonded alloy skins, the whole being locked into position by three toggles on either side. Similar waffle construction was employed on parts of the engine cowl, whose four hinged panels gave almost unrestricted access to the BMW engine. Further access to the engine accessories was afforded by hinged panels aft of the exhaust stubs on either side of the fuselage. Each panel had three louvres for the egress of engine cooling air and on the Fw 190A-4 and later models these louvres were adjustable by the pilot.

The cockpit cover and its fairing was built up as an integral unit, with the base of the structure consisting of a $\frac{5}{8}$ in (15,8 mm) tubular member bent into an inverted-U in the front to fit into the windshield. The Plexiglass of the canopy was mounted between two strips of buna and a flat aluminium strip, held by screws driven into self-locking nuts in the tube. At the rear of the canopy a stamped aluminium A-frame was set between the tube-frame ends, and was riveted to an alloy fairing mounted on a $\frac{3}{4}$ in (19 mm) tube extending aft. The whole structure ran on three ball-bearing rollers, one on each side of the canopy in the top-fuselage longerons and one attached to the $\frac{3}{4}$ in (19 mm) tube, running in a channel section set in the fuselage turtle deck. The pilot wound the canopy open and closed by a ratchet-and-sprocket-operated handle on the right of the cockpit. In an emergency the whole canopy could be jettisoned by a small lever placed near the operating handle.

Depression of the canopy jettison lever

A later acquisition at the RAE was this Fw 190A-4/U-8 *Jabo*, tested as PE882. In the photograph above, the wing pylons are still fitted; in the illustration at the top of the opposite page, they have been removed

first disengaged the canopy winding sprocket and secondly released, via a series of rods and shafts, the latch of a spring-loaded firing pin; the pin fired a standard 20-mm cartridge which physically sheared the shaft inserted in the canopy tube and blew the entire canopy sufficiently far back for the slipstream to lift it away. The canopy carried with it the 14-mm (0.55-in) armour headplate and broke the FuG 7 aerial. During ground maintenance a safety latch was inserted into the firing block.

Tail Unit
The aft fuselage, which extended from bulkhead No 14, consisted of a integral fuselage and tail fin and contained the bell-cranks and push-rods for the elevator and rudder, the stabiliser trim motor and raising and lowering mechanism for the tailwheel. The entire tail unit was attached to bulkhead No 14 by flanged mating bulkheads and a series of closely spaced bolts, while two ribs, mounted horizontally and $7\frac{3}{8}$ in (18,7 cm) apart, carried the adjustable tailplane. Both ribs intersected a diagonal member which was the main stress member for the tail unit and carried tailwheel loads when on the ground and the aerodynamic loads while airborne. This member started at the bottom skin and $18\frac{3}{4}$ in (47,6 cm) aft of the bulkhead and extended diagonally up and aft $63\frac{1}{2}$ in (1,61 m) to the top of the vertical fin rib. The tailwheel drag yoke was riveted to the diagonal member 9 in (23 cm) from its base, while a forged hexagonal fitting was riveted between the ribs already mentioned, and the elevator spar was attached to it. On the aft face of the diagonal member the guide rails for the tailwheel retracing unit were mounted. The topmost of the horizontal ribs extended aft to pick up the centre hinge of the rudder, while the lower rib ran down 28 deg to support the lower rudder hinge. Further ribs and Z-section stringers lent strength to the tail unit, while the construction aft of the diagonal member and above the elevator was of the waffle type, obviating the further use of stringers; a triangular inspection door 30 in (76 cm) high with a 15 in (38 cm) base was set in the left side of the tail fin and gave easy access to the tailwheel operating mechanism.

The dynamic and mass-balanced rudder was built around a single spar of stamped, flanged aluminium to which were riveted the three hinge fittings, with the leading edges flush-riveted to the spar and the ribs made up of rounded gusset plates. The rudder was fabric-covered and the rudder trim, which

Above, Fw 190A–4/U8 PE882 (see opposite page). Below, an Fw 190A–5/U14 torpedo carrier, as indicated by the lengthened tail wheel leg

Below, an Fw 190A–5/U8, *Werk-Nr* 2569 of I/SKG 10. This example landed in error at RAF Manston on 22 June 1943 and as indicated by the black undersides it had been used for night operations

135

Weights and Performance
All following data relate to the Focke-Wulf Fw 190A–3, *Werk-Nr* 5313.

Weight Summary

Empty weight with all fixed equipment except weapons	6,544 lb	(2 970 kg)
Pilot and parachute	200 lb	(90 kg)
Fuel, 523 litres (115 gallons)	860 lb	(390 kg)
Oil, 45 litres (10 gallons)	96 lb	(43,5 kg)
Two MG 121/20 cannon	196 lb	(89 kg)
Ammunition for above (2 × 200 rounds)	200 lb	(90 kg)
Two MG FF cannon	126 lb	(57 kg)
Ammunition and magazines for above (2 × 55 rounds)	90 lb	(41 kg)
Two MG 17 machine guns	56 lb	(25 kg)
Ammunition for above (2 × 1,000 rounds)	142 lb	(64 kg)
Radio equipment	70 lb	(31 kg)
Loaded Weight	8,580 lb	(3 890 kg)

Weight during trials	8,580 lb	(3 890 kg)
Power loading	5.36 lb/bhp	(2,43 kg/hp)
Wing loading	42.3 lb/sq ft	(206 kg/m²)

Maximum level speeds

Height	Speed (TAS)	RPM	Boost	Remarks
4,500 ft (1 372 m)	327 mph (526 km/h)	2,450	1.35 atas (19.8 lb/sq in)	M Gear, full throttle height
8,000 ft (2 440 m)	317 mph (510 km/h)	2,450	1.21 atas (17.8 lb/sq in)	M Gear
18,000 ft (5 500 m)	375 mph (603 km/h)	2,450*	1.35 atas (19.8 lb/sq in)	S Gear, full throttle height
20,000 ft (6 100 m)	369.5 mph (594 km/h)	2,450	1.27 atas (18.7 lb/sq in)	S Gear
25,000 ft (7 625 m)	351 mph (565 km/h)	2,450	1.07 atas (15.7 lb/sq in)	S Gear

*Full output of the BMW 801D-2 at rated height of 18,000 ft (5 500 m) was assessed as 1,455 bhp, which gave the Fw 190A–3 a maximum TAS of 391 mph (629 km/h) and the above figures relate to the de-rated version only

Rates of climb

Height	Rate of climb	RPM	Boost	Remarks
0–4,000 ft (1 220 m)	2,900 ft/min (14,7 m/sec)	2,350	1.28 atas (18.8 lb/sq in)	30-minute limit
0–4,000 ft (1 220 m)	3,250 ft/min (16,5 m/sec)	2,450	1.35 atas (19.8 lb/sq in)	3-minute limit
8,000 ft (2 440 m)	2,200 ft/min (11,2 m/sec)	2,350	1.10 atas (16.2 lb/sq in)	M Gear
8,000 ft (2 440 m)	2,450 ft/min (12,5 m/sec)	2,450	1.17 atas (17.2 lb/sq in)	
10,000–17,500 ft 3 050–5 337 m)	2,800 ft/min (14,2 m/sec)	2,350	1.28 atas (18.8 lb/sq in)	S Gear. 30-minute limit
10,000–17,500 ft 3 050–5 337 m)	3,500 ft/min (17,8 m/sec)	2,450	1.35 atas (19.8 lb/sq in)	S Gear. 3-minute limit
25,000 ft (7 625 m)	2,000 ft/min (10,2 m/sec)	2,450	1.03 atas (15.1 lb/sq in)	S Gear

Service Ceiling:
  32,700 ft (9 975 m) (assessed as height at which rate of climb reduces to below 500 ft/min 2.54 m/sec)

Operating Speeds and Limits

Take-off	93–100 mph (150–160 km/hr) IAS
Take-off run (nil wind)	1,150–1,320 ft (350–400 m)
Max undercarriage limit	155 mph (250 km/hr) IAS
Max rate of climb speed	174 mph (280 km/hr) IAS
Diving speed limit (VNE)	466 mph (750 km/hr) IAS
Approach speed	124–137 mph (200–220 km/hr) IAS
Landing speed	112 mph (180 km/hr) IAS

Range and Endurance

At 1,000 ft (305 m) altitude	1 hr 5 min	: 332 mls (535 km)
At 16,400 ft (5 000 m) altitude	1 hr	: 324 mls (520 km)
At 29,500 ft (9 000 m) altitude	1 hr 25 min	: 420 mls (675 km)

was adjustable only on the ground, consisted of a 15 in by 1 in (38 cm by 2,5 cm) perforated metal strip fixed to the trailing edge. The elevator was a single spar all-metal cantilever surface with seven floating ribs on each side; except for the detachable tips, the unit was in one piece.

The electric trimming for the stabiliser in the Fw 190 was a refinement not encountered in its contemporaries. An electric motor was suspended by a ball and socket joint from the leading edge of the vertical fin and was attached to the leading edge of the elevator by means of a screw-jack and yoke. The motor operated at 14,000 rpm and incorporated six trains of gears giving a 533 to 1 reduction that moved the leading edge at a rate of 4.1 in (10,4 cm) per minute or over the full range from 3 deg nose *up* to 5 deg nose *down* in about 20 seconds. The pilot operated the electric trimmer by pressing either one of two buttons on the left-hand console and, in addition, had an electric trim guage. Normal setting for take-off was 0 deg and great care had to be exercised when using the powerful trimmer at high speed and low level, for it injudicious use could accelerate a high-speed stall and subsequent 'flick'. The elevators were built around a single spar, with metal leading edge, ribs and trailing edge, the whole unit being fabric covered with adjustable metal trim tabs in the same manner as the rudder.

Ancillary Components

Undercarriage

The undercarriage of the Fw 190 was unusually high and wide, rendering the aircraft particularly suited to the carriage of bulky ordnance under the fuselage and to good ground handling. Each wheel, fitted with a 700 × 175 mm smooth-tread tyre, was mounted at the end of a single long oleo leg with a concertina cover and triangular torsion links, the oleo having a possible travel of 15 in (38 cm). Each leg was hinged at the top on a shaft that had bearings in the light leading edge and in the main spar. Each leg was raised or levered by an electric motor mounted on the back of the main spar web, which ran at 14,000 rpm and was geared down to 10,494:1. When retracted, electrically operated up-locks with a mechanical override held the gear in place, but when it was extended, only the mechanical lock formed by the side stays and the high inertia of the motor sufficed to lock the undercarriage in the 'down' position. Braking of the wheels was via hydraulically actuated units mounted on the rudder pedals. Tyre pressure for the main wheels was 4 atas (58.8 lb/sq in).

The tailwheel retraction was actuated by retracting the right undercarriage by means of pulleys and cables. The upper end of the tailwheel oleo piston carried a toggle with two rollers placed inside a channel mounted to the rear of the fin diagonal member – so shaped that when the rollers were at their lowest point, they were locked thus locking the tailwheel in the down position. When the main undercarriage started to retract,

A close-up of the cowling, undercarriage and cockpit of an anonymous Fw 190 tested in the USA

tension was placed on the cable, which then broke the lock by pulling the rollers out of their recess and pulled the tailwheel unit up its channel against the tension of a spring. The tailwheel, size 350 × 135 mm, was mounted in a steel fork which was rotatable through 360 deg but was so mounted that it castored freely while being centred by a hairpin spring; when the control column was pulled fully back, the tailwheel was locked in the central position. Tyre pressure was 4.5 *atas* (66 lb/sq in).

Weapons and Armour
Standard armament for the Fw 190A–3, as carried by the captured *Werk-Nr* 5313, comprosed two Rheinmetall MG 17 machine guns of 7,92-mm calibre fitted in Station 1 (central fuselage position), two Mauser MG 151/20 cannon of 20-mm calibre in Station 2 (inner wing) and two Rheinmetall MG FF 20-mm cannon in Station 3 (outer wing). The MG FF cannon were optional and their carriage depended on the nature of air fighting that prevailed.

The MG 17 machine guns were standard for all types of Fw 190 from the A–1 up to and including the A–6 of 1943, and these performed the useful task of 'sighter' weapons when using tracer ammunition. The cocking and firing of the MG 17 were respectively pneumatic and electric, with an air bottle installed in the mounting providing air at a pressure of 20 *atas* (300 lb/sq in). The pilot cocked the weapon by depressing a button on the KG 13A control column and fired the same by selecting either Position 1 or 3 and pulling the trigger. Ammunition in metal disintegrating belts was fed from two magazines, each of 1,000 rounds capacity, to the front of bulkhead No 1 and gave the fuselage weapons about 60 seconds' firing time, while 'empties' were expended to atmosphere. The muzzle velocity of the MG 17 when using Ball ammunition was 2,800 ft (850 m) per sec with a rate of fire of 1,000 rounds per minute, controlled by the synchronising unit mounted on the rear of the BMW 801.

Cocking and firing of the Mauser MG 151/20 cannon was electric throughout, necessitating the use of special 20-mm ammunition for synchronous firing. The interrupter gear – a DSG 3 AL unit, mounted on the engine – passed 24–volt current to the Mauser for firing, while interruption for the MG 17 machine-guns was mechanical. Light alloy panniers, mounted between the main spar and the front fuel tank, held a maximum of 250 rounds per gun (only 200 were normally carried) and with a rate of fire of 700 rounds per minute gave approximately 17 seconds' firing time. The weapon fired various types of round but usually ammunition was fired in the following sequence: armour-piercing (AP), high-explosive/incendiary (HE/I), armour-piercing/high-explosive (AP/TE) with every eighth round an incendiary/tracer (I/T) to assist in gunlaying. The muzzle velocity of the MG 151/20 cannon varied from 2,596 ft (790 m) per sec with HE ammunition to 2,364 ft (720 m) per sec with armour-piercing.

The outer wing cannon in Station 3 were the old well-tried Rheinmetall MG FF guns of the Oerlikon type. The weapon was cocked electro-pneumatically by a solenoid and associated air bottle in the cannon bay, while firing was operated electrically. A cannular magazine held 60 rounds of 20-mm ammunition of AP, HE and I/T rounds sufficient for $7\frac{1}{2}$ seconds' firing time with a rate of fire of 520 rounds per minute; muzzle velocity was low at 1,920 ft (585 m) per sec.

The pilot was provided with a Carl Zeiss Revi C 12/D reflector sight, a SZKK 4 rounds-counter and a control unit which indicated cocking of the MG 17 and MG 151 guns by lights and gave an immediate indication of the remaining ammunition; an additional rounds counter was provided for the MG FF, when fitted. A selector switch on the left of the pistol grip gave the pilot a

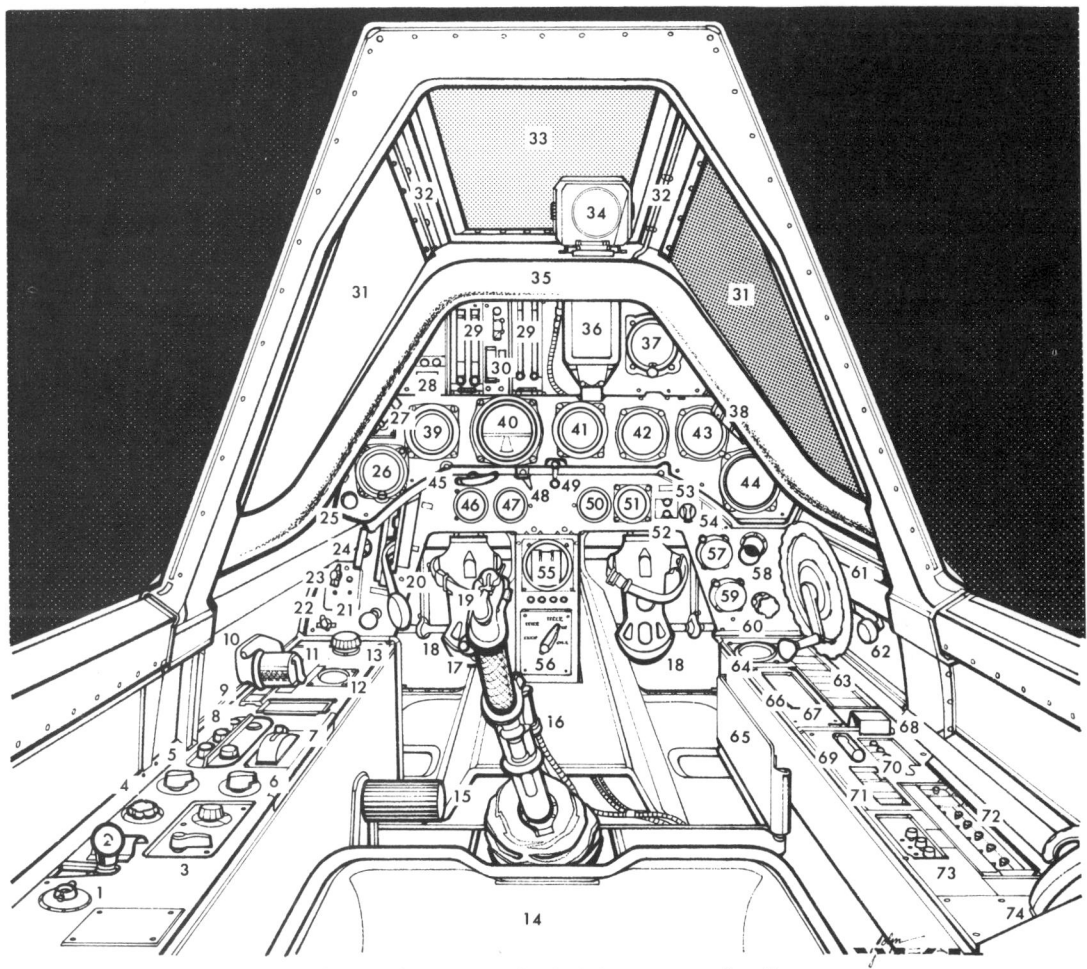

Focke-Wulf 190A-8 Cockpit Instrumentation Key

1. Helmet R/T connection
2. Primer fuel pump handle
3. FuG 16ZY communications and homing switch and volume control
4. FuG 16ZY receiver fine tuning
5. FuG 16ZY homing range switch
6. FuG 16ZY frequency selector switch
7. Tailplane trim switch
8. Undercarriage and landing flap actuation buttons
9. Undercarriage and landing flap position indicators
10. Throttle
11. Throttle-mounted propeller pitch control thumbswitch
12. Tailplane trim indicator
13. Instrument panel lighting dimmer
14. Pilot's seat
15. Throttle friction knob
16. Control column
17. Bomb release button
18. Rudder pedals
19. Wing gun firing button
20. Fuel tank selector lever
21. Engine starter brushes withdrawal button
22. Stop cock control lever
23. FuG 25a IFF control panel
24. Undercarriage manual lowering handle
25. Cockpit ventilation knob
26. Altimeter
27. Pitot tube heater light
28. MG 131 "armed" indicator lights
29. Ammunition counters
30. SZKK 4 armament switch and control panel
31. 30-mm armour glass quarter lights
32. Windscreen spray pipes
33. 50-mm armour glass
34. Revi 16B reflector gunsight
35. Padded coaming
36. Gunsight padded mounting
37. AFN 2 homing indicator (FuG 16ZY)
38. Ultra-violet lights (port and starboard)
39. Airspeed indicator
40. Artificial horizon
41. Rate of climb/descent indicator
42. Repeater compass
43. Supercharger pressure gauge
44. Tachometer
45. Ventral stores manual release
46. Fuel and oil pressure gauge
47. Oil temperature gauge
48. Windscreen washer operating lever
49. Engine ventilation flap control lever
50. Fuel contents gauge
51. Propeller pitch indicator
52. Rear fuel tank switchover light (white)
53. Fuel contents warning light (red)
54. Fuel gauge selector switch
55. Underwing rocket (WGr 21) control panel
56. Bomb fusing selector panel and (above) external stores indicator lights
57. Oxygen flow indicator
58. Flare pistol holder
59. Oxygen pressure gauge
60. Oxygen flow valve
61. Canopy actuator drive
62. Canopy jettison lever
63. Circuit breaker panel cover
64. Clock
65. Map/chart holder
66. Operations data card
67. Flare box cover
68. Starter switch
69. Flare box cover plate release knob
70. Fuel pump circuit breakers
71. Compass deviation card
72. Circuit breaker panel cover
73. Armament circuit breakers
74. Oxygen supply

139

An Fw 190A–5 captured before the end of the war and taken to the USA for evaluation at Wright Field. Unauthentic *Luftwaffe* markings were subsequently restored, as shown in this illustration

choice of weapons with Position 1 firing MG 17 and MG 151/20 guns, Position 2 firing the MG FF cannons only and Position 3 firing all guns. In addition, circuit breakers (fuses) located on the righthand console gave further control over individual pairs of guns or combinations of pairs.

The harmonisation found on the Fw 190A-3 *Werk-Nr* 5313 was with the twin MG 17s ranged at 330 yd (300 m), the twin MG 151/20s at 490 yd (450 m), and the twin MG FFs at 270 yd (245 m), giving a concentration of fire with the smallest bullet grouping at about 270 yd (245 m).

Armour protection for the Fw 190 pilot was heavy, with the bulk of the engine and a sloping 50-mm (1.97-in) armourglass windshield affording cover from the front. From behind, protection was afforded by a 14-mm (0.55-in) head and shoulder plate bolted to the sliding canopy, an 8-mm (0.31-in) plated seat back and four small plates 8 mm (0.31 in) thick, placed around the rear of the seat. The oil tank and radiator were protected by a 3-mm (0.12-in) front cowl piece, and this was further reinforced by a 5 mm (0.197 in) ring that fitted closely around the cooling fan.

Bombs or drop tanks were carried on the electrically operated ETC 501/XIIA carrier, which was itself mounted in lugs at the base of fuselage main spar and bulkhead No 5. The fuse junction box was placed in the fuselage behind the cockpit area and a fuse selector in the cockpit centre panel.

Radio Equipment

Until the installation of FuG 16Z in the Fw 190A-4, all Focke-Wulfs had the FuG 7 *(Funk Gerät)* type R/T made by the Lorenz concern. The equipment consisted of a transmitter and receiver box, motor generator, junction box/power pack, selector box with volume control and station selector and send/receive button on the pilot's control column. Powered from the 24–volt aircraft mains, the set operated within the range of 2.5–3.75 MHz frequencies of the 80–120-m band and had a performance roughly equivalent to the RAF's TR 9 HF radio. The later FuG 16Z operated in VHF wavebands. For radar interpretation of 'hostile' or 'friendly' aircraft, FuG 25 IFF equipment was carried.

Powerplant

The powerplant fitted in the Fw 190A-3 fighter, and subsequent series up to and including the Fw 190A-8, was the BMW 801D-2 engine manufactured by the Bayerische Motoren Werke Flugmotorenbau GmbH of Munich. This engine was a compact 14-cylinder radial fitted with a single-stage two-speed supercharger, giving a rated output of 1,770 bhp for take-off and emergency combat. The double rows of seven

Another view of the Fw 190A–5 tested at Wright Field, also shown on the opposite page. The original code letters can still be discerned on the fuselage and under the wing, but the crosses have been repainted

cylinders each were air-cooled and fitted with direct fuel injection, which maintained fuel supply to the cylinders under negative G conditions.

The bore and stroke of the BMW 801D–2 was 156 × 156 mm (6.15 × 6.15 in), giving a total swept capacity of 42 litres (2 563 cu in), with each cylinder having one inlet valve, one exhaust valve, two spark plugs and a single fuel injector. The inlet valves were made of aluminium-bronze while the exhaust valves were steel filled with sodium; at the very high temperatures experienced by the exhaust valves, the sodium melted to liquid form and helped to dissipate much of the heat. The connecting rods from piston to crankshaft consisted of steel H-form rods having lead-bronze big-end bearings with one master and six auxiliaries per bank of cylinders driving a two-throw four-piece crankshaft mounted on a single-row ball bearing and two roller-bearings set in the forward and rear sections of the steel crankcase. Bolted to the front portion of the crankcase was a light alloy housing which enclosed the airscrew reduction gears, the cooling fan increasing gears, the magneto drives and the airscrew pitch adjusting mechanism. Mounted on the rear of the crankcase was the blower casing, consisting of two parts: the forward section housed the front bearing of the supercharger, the centrifugal impeller and the baffles while individual pipes routed the air-fuel mixture forward to each cylinder; and the rear portion housed the supercharger drive, the rear bearing and the various accessory drives.

The supercharger consisted of a two-speed centrifugal blower with the light-alloy 24-blade impeller fed with air delivered by the cooling fan via shallow ducts on either side of the engine cowling, which by-passed the cooling air for the cylinders. The supercharger drive unit from the crankshaft con-

BMW 801D–2 Power Ratings (unrestricted)

Rating	Supercharger	RPM	Boost (ata)	Power (bhp)	Height
Take-off	'M' Low gear	2,700	1.42	1,695	MSL
Max emergercy	'M' Low gear	2,700	1.42	1,755	3,250 ft (990 m)
Climb & 30 min	'M' Low gear	2,450	1.35	1,595	2,500 ft (760 m)
Climb & 30 min	'S' High gear	2,450	1.35	1,455	18,000 ft (5 500 m)
Combat	'S' High gear	2,700	1.42	1,530	20,000 ft (6 100 m)

(These figures were taken from an operational Focke-Wulf 190A–4/U8 fighter-bomber of I/SKG 10 in June 1943 and are representative of the outputs of a standard BMW 801D–2 radial engine)

The fine lines of the closely-cowled BMW 801D in the Fw 190 are well shown in this illustration of an early production example

sisted of two freely mounted gears fitted with multiple plate clutches and a sliding selective clutch member which was hydraulically operated from the master control unit *(Kommandgerät)*. The two-speed gearing was made up of High Gear (S) which drove the impeller directly from the crankshaft pinion to a forward gear mounted on the output shaft at a ratio of 7.46 : 1, while Low Gear (M) was driven via two free gears, one of which was clutched, to the impeller at a ratio 5.07 : 1. The changeover altitude in the Focke-Wulf 190 from M to S Gear was set automatically at 8,200 ft (2 500 m) and there was no way in which the pilot could override the automatic changeover. An example of how this affected the performance is given by the rate of climb with maximum power selected at 2,450 rpm and 1,35 atas boost: rate of climb up to 4,000 ft (1 220 m) remained at approximately 3,050 ft/min (15,5 m/sec), dropping to 2,250 ft/min (11,44 m/sec) at full throttle height in M Gear just below 8,000 ft (2 400 m) until the changeover to S Gear, which restored the rate of climb to about 3,280 ft/min (16,68 m/sec) at 17,500 ft (5 350 m) at the Fw 190's rated altitude, where the power delivered from the engine represented its maximum output at height. Above the rated altitude power fell off in the same manner as in an unsupercharged engine.

A Deckel fuel injection pump provided C 3 (100 Octane) petrol via a de-aerator and high-pressure lines to the injector nozzles, placed between the valves, in each cylinder. The pump was mounted, along with other accessories, on the rear blower housing and was driven at 0.167 of the crankshaft speed via gears from the shaft. Ignition was effected by two Bosch DWT 240 ET 7 high-altitude plugs per cylinder, served by a twin Bosch ZM 14 CR–10 magneto mounted vertically on top of the alloy front casing and driven via gears from the forward crankshaft.

The airscrew drive was reduced by an epicyclic reduction gear via the hollow crankshaft to a gear ratio of 0.54 : 1, and an external gear on the casing of the epicyclic unit drove the cooling fan through a layshaft at 1.72 times the engine speed. The airscrew pitch-change was by hydraulic pump mounted on the engine nose casing, with the pump being controlled by an electric motor, itself monitored by the *Kommandgerät*, which provided constant-speed, variable pitch and overspeed protection. The propeller was a VDM type made of light metal with a diameter of 10 ft 11 in (3,34 m) and a fairly high solidity of 0.13; the VDM propeller was controlled either automatically, or manually by a selector switch and thumb-switch on the pilot's throttle.

Lubrication for the BMW 801D–2 consisted of an annular tank mounted in the nose cowl with a capacity of approximately 55–8 litres (12 Imp gal) of *Rotring* 491 oil, which was fed, under a pressure of between

8 to 9 atas (118–32 psi), via an auxiliary and main pressure pump, oil cooler and filter to various parts of the engine. A relief valve limited pressure through the oil cooler to a maximum of 12.5 *atas* (184 psi). The return of oil from the crankcase, supercharger and accessory mounting was from an oil sump via the main return pump; two additional scavenge pumps drained the gear casing and a further pump drained the rocker boxes. A small separate oil tank at the rear of the engine supplied pressure oil for lubrication of the Deckel injector pump and boost control oil servo controls in the *Kommandgerät*. For normal flying conditions oil temperatures were 70 deg C maximum at the inlet and 105 deg maximum at the outlet.

The cooling of the engine was by air, and airflow through the engine cowl was directly proportional to forward speed and the speed of the cooling fan. The 12-bladed fan was 32 in (81 cm) in diameter and was driven at 3.17 times the speed of the propeller. Compressed fresh air from behind the pressure region created by the fan was routed through ornate light-alloy baffles which closely enclosed the individual cylinders and their heads. Cooling had been a constant problem in the BMW 801 motors and cooling air passing through the first row of cylinders was carefully scavenged by baffles to feed the rear row; to avoid cooling loss within the engine cowling, the cylinder-head baffles were extended to form a dividing wall between the two rows of cylinders and were sealed to the cowling by a rubber ring. Cooling air for the annular oil cooler passed through the fan, reversed direction and passed through an 18 mm (0.70 in) annular gap between the front armour ring and the front portion of the cowling. This slot could not be adjusted in the BMW 801C and D series of engines.

The exhaust pipes were about 3 in by 1.5

Above right, accessibility of the BMW 801 engine was excellent, through use of large hinged panels. Below, a head-on view of the captured Fw 190A–4/U8 PE882

143

in (7,6 cm by 3,8 cm) in cross section and were distributed about the rear cowl, with four on either side and six underneath; two of the latter were mounted slightly outboard and forward of the remainder. Muff-type heaters were placed about some of the exhaust pipes and hot air was ducted by pipes to the cockpit, the ammunition chutes for the MG 17 machine guns and the outboard 20-mm MG FF cannon, when installed.

Among the accessories mounted on the rear casing of the blower was a Bosch 24-volt 2 kilowatt generator, an Ascania vacuum pump, a Maihak high-pressure fuel pump and the Deckel injection pump. A starter unit was mounted on the rear casing, with the extended end of the crankshaft passing through the hollow impeller shaft to the starter dogs and auxiliary gear drives of a co-axially mounted Bosch electric or hand-energised inertia starter. The Fw 190 was usually started by plugging in an external trolley-accumulator but could be started by manual cranking or from the internal battery, although this procedure was not recommended for normal use.

Mounted on the right-hand side of the rear casing was the complicated electro-hydraulic master control (*Kommandgerät*), connected to a single lever on the cockpit console. When the pilot moved the 'throttle', the movement was transmitted by bell-cranks and push-pull rods to the *Kommandgerät* which, in turn, controlled fuel flow, fuel mixture, propeller pitch setting, ignition timing and automatic change of gear for the supercharger.

Hydraulic oil for the *Kommandgerät* was held in the engine mounting ring (capacity 5-6 litres, 1.2 Imp gal), which was bolted to a light-alloy channel-section ring on the rear of the blower casing by ten rubber-bushed mounting blocks. Welded-steel tube engine mounting rods were bolted to the mounting ring at four points, and to the firewall (bulkhead No 1) by five spherical joints. Under field conditions the whole 'power egg' could be removed and changed within four hours by as few as three men.

Fw 190 Comparative Dimensions

Variant	Span	Length	Wing area
Fw 190 V1	31 ft 2½ in (9,515 m)	29 ft 0½ in (8,850 m)	160.38 sq ft (14,9 m²)
Fw 190 V5k	31 ft 2 in (9,500 m)	28 ft 10½ in (8,789 m)	160.38 sq ft (14,9 m²)
Fw 190 V5g	34 ft 0¾ in (10,383 m)	28 ft 10½ in (8,789 m)	196.98 sq ft (18,3 m²)
Fw 190A (to A–4)	34 ft 0¾ in (10,383 m)	28 ft 10½ in (8,789 m)	196.98 sq ft (18,3 m²)
Fw 190A–5	34 ft 0¾ in (10,383 m)	29 ft 4¼ in (8,950 m)	196.98 sq ft (18,3 m²)
Fw 190A–6 (and subsequent)	34 ft 5½ in (10,506 m)	29 ft 4¼ in (8,950 m)	196.98 sq ft (18,3 m²)
Fw 190A–10	37 ft 8¾ in (11,500 m)	29 ft 4¼ in (8,950 m)	220.66 sq ft (20,5 m²)
Fw 190B	40 ft 4¼ in (12,300 m)†	29 ft 4¼ in (8,950 m)	218.51 sq ft (20,3 m²)
Fw 190C	34 ft 5½ in (10,506 m)	30 ft 10¾ in (9,420 m)*	196.98 sq ft (18,3 m²)
Fw 190D–1 (V17)	40 ft 4¼ in (12,300 m)	33 ft 5¼ in (10,192 m)	218.51 sq ft (20,3 m²)
Fw 190D–9	34 ft 5½ in (10,506 m)	33 ft 5¼ in (10,192 m)	196.98 sq ft (18,3 m²)
Ta 152B	36 ft 1 in (11,000 m)	35 ft 1½ in (10,710 m)	209.89 sq ft (19,5 m²)
Ta 152H	47 ft 4½ in (14,440 m)	35 ft 1½ in (10,710 m)	250.80 sq ft (23,3 m²)

† Tested on Fw 190B–1 *Werk-Nr* 0046, alternative to 34 ft 5½ in (10 506 m).

* Prototype lengths varied, up to 33 ft 5¼ in (10,192 m) for Fw 190 V32/U1.

Development Batch, Prototypes, Production Variants

Fw 190 Prototypes

Designation	Werk-Nr	Engine	Notes
Fw 190 V1	0001	BMW 139	First flown 1–6–39. NACA cowling later replaced ducted spinner
Fw 190 V2	0002	BMW 139	Similar to V1, first flown 1–12–39. NACA cowling fitted later
Fw 190 V3	0003	(BMW 139)	Not flown. Cannibalised for Fw 190 V1 & V2
Fw 190 V4	0004	(BMW 139)	Not flown. Used for static tests
Fw 190 V5	0005	BMW 801 C–0	Revised wing, in two forms – short (k) and long (g) span. Crashed
Fw 190 V6	0006	BMW 801 C–0	First Fw 190A–0. Short-span wing. Became Fw 190A–0/U1
Fw 190 V7	190.0110–001	BMW 801 C–1	First Fw 190A–1
Fw 190 V8	0021	BMW 801 C–0	Proposed designation for Fw 190A–0/U3 (tests of FuG 25 IFF). Crashed 1–10–41
Fw 190 V9	0022–23	BMW 801 C–0	Proposed designation for two Fw 190A–0 trials aircraft, not used
Fw 190 V10	0015	BMW 801 C–1	Proposed designation for Fw 190A–0/U5
Fw 190 V11	—	Wright	Proposed installation of Wright Cyclone engine, not built
Fw 190 V12	0035	BMW 801	Pressure cabin in Fw 190A–0
Fw 190 V13	0036	DB 603A	First DB 603 installation, in Fw 190A–0. Crashed 30–7–42
Fw 190 V14	190.0120–201	BMW 801 C–2	First Fw 190A–2
Fw 190 V15	0037	DB 603A	Prototype for Fw 190C
Fw 190 V16	0038	DB 603A	Prototype for Fw 190C
Fw 190 V17	0039	Jumo 213	First Jumo installation, in Fw 190A–0. Pressure cabin. Increased span later. Became Fw 190 V17/U1
Fw 190 V17/U1	0039	Jumo 213A	Prototype for Fw 190D–9, May 1944
Fw 190 V18	0040	DB 603A	Prototype for Fw 190C, became Fw 190 V18/U1

Fw 190 V18/U1	0040	DB 603A	Unpressurised. Hirth exhaust-driven turbo compressor in ventral fairing. 'Känguruh'. Became V18/U2
Fw 190 V18/U2	0040	DB 603A	Converted to prototype Ta 152H–1. Unpressurised
Fw 190 V19	0041	Jumo 213A	Revised wing, longer fuselage. Unpressurised. Prototype for Fw 190D Series. Crashed 16-2-44
Fw 190 V20	0042	Jumo 213A	Similar to V19. Became 190 V20/U1
Fw 190 V20/U1	0042	DB 603L	Destroyed 5-8-44 while under conversion to Ta 152C configuration
Fw 190 V21	0043	Jumo 213A	Similar to V19. Became Fw 190 V21/U1
Fw 190 V21/U1	0043	DB 603E	Converted November 1944 as D–B test-bed
Fw 190 V22	0044	Jumo 213C	Prototype Fw 190D Series. Pressurised
Fw 190 V23	0045	Jumo 213C	As for Fw 190 V22
Fw 190 V24	190.0140–561	BMW 801D	First Fw 190A–4
Fw 190 V25	0050	Jumo 213A	Similar to V19. Prototype for Fw 190D–1
Fw 190 V26	0051	Jumo 213A	Similar to V19 but pressurised. Prototype for Fw 190D–2
Fw 190 V27	0052	Jumo 213A	Similar to V26
Fw 190 V28	0053	Jumo 213A	Similar to V19. Used for static tests
Fw 190 V29	0054	DB 603A or S	Similar to V18/U1 but pressurised. 'Känguruh'. Became Fw 190 V29/U1
Fw 190 V29/U1	0054	Jumo 213E	Converted to prototype Ta 152H
Fw 190 V30	0055	DB 603A or S	Similar to V29. Became Fw 190 V30/U1
Fw 190 V30/U1	0055	Jumo 213E	Converted to prototype Ta 152H. Crashed 13-8-44
Fw 190 V31	0056	DB 603A or S	Similar to V29. Crashed 29-5-43
Fw 190 V32	0057	Jumo 213E	Similar to V29. Long-span wings fitted. Became Fw 190 V32/U1
Fw 190 V32/U1	0057	DB 603G	Converted to serve as D–B test-bed in December 1943. Became Fw 190 V32/U2
Fw 190 V32/U2	0057	DB 603A or S	Converted to prototype Ta 152H–1, with wings of Ta 152 V25, in March 1945
Fw 190 V33	0058	Jumo 213E	Similar to V29. Converted to Fw 190 V33/U1
Fw 190 V33/U1	0058	DB 603A or S	Converted to prototype Ta 152H. Crashed 23-7-44

Fw 190 V34	410230	BMW 801F-1	Prototype Fw 190A-9. Engine test-bed
Fw 190 V35	816	BMW 801F-1	Prototype Fw 190A-9. BMW 801TU test-bed
Fw 190 V36	—	BMW 801F-1	Prototype Fw 190A-9
Fw 190 V37	—	BMW 801D-2	Fw 190A-5 weapons trials, four MG 151 in outer wing bays
Fw 190 V38	—	BMW 801D-2	As V37, four MK 108s in outer wing bays
Fw 190 V39	—	BMW 801D-2	As V37, two MK 103 under wings
Fw 190 V40	—	—	No information
Fw 190 V41	—	—	No information
Fw 190 V42	—	BMW 801D-2	Fw 190A-5/U2, special trials
Fw 190 V43	—	—	No information
Fw 190 V44	—	—	No information
Fw 190 V45	7347	BMW 801D-2	Prototype Fw 190A-6/R4. GM 1 booster trials
Fw 190 V46	—	Jumo 213C	Fw 190D-series prototype, similar to Fw 190 V17
Fw 190 V47	530115	BMW 801D-2	Similar to Fw 190 V45
Fw 190 V48	—	—	No information
Fw 190 V49	—	—	No information
Fw 190 V50	—	—	No information
Fw 190 V51	530766	—	Fw 190A-5 weapons trials, two MK 108s in outer wing bays
Fw 190 V52	—	—	No information
Fw 190 V53	170003	Jumo 213A	Prototype for Fw 190D-9, similar to V17/U1
Fw 190 V54	174024	Jumo 213A	As V55. Destroyed 5-8-44
Fw 190 V55	170923	Jumo 213E(F)	Prototype for Fw 190D-11
Fw 190 V56	170924	Jumo 213E(F)	As V55
Fw 190 V57	170926	Jumo 213E(F)	Modified Fw 190A-8 as Fw 190D-11
Fw 190 V58	170933	Jumo 213E(F)	As V57
Fw 190 V59	350156	Jumo 213E(F)	As V57
Fw 190 V60	350157	Jumo 213E(F)	As V57
Fw 190 V61	350158	Jumo 213E(F)	As V57
Fw 190 V62	732053	Jumo 213E	Prototype Fw 190D-13, *Motorkannon*
Fw 190 V63	350166	Jumo 213E	Modified Fw 190A-8 as Fw 190D-12
Fw 190 V64	732054	Jumo 213E	As V63
Fw 190 V65	350165	Jumo 213E	As V63
Fw 190 V66	584002	BMW 801TS/TH	Re-engined Fw 190A-8 as Fw 190F-15
Fw 190 V67	930516	BMW 801TS/TH	Prototype for Fw 190F-16, May 1945

Fw 190 V68	170003	Jumo 213A	Fw 190 V53 rebuilt with MK 103s in wing roots
Fw 190 V69	582092	BMW 801D-2	Fw 190A-8 for trials with X-4
Fw 190 V70	580029	BMW 801D-2	As V69. Crashed 25-8-44
Fw 190 V71	350167	Jumo 213E	As V63
Fw 190 V72	170727	BMW 801TS	Prototype for Fw 190A-9
Fw 190 V73	733705	BMW 801TS	Prototype for Fw 190A-9
Fw 190 V74	733713	BMW 801TS	Prototype for Fw 190A-9. SG 117 weapon trials
Fw 190 V75	582071	BMW 801D	Fw 190F-8 with SG 113A weapons
Fw 190 V76	210040	DB 603LA	Fw 190D-9 modified to D-14 prototype
Fw 190 V77	210043	DB 603E	Similar to Fw 190 V76
Fw 190 V78	—	BMW 801D-2	No information
Fw 190 V79	—	BMW 801D-2	No information
Fw 190 V80	—	BMW 801D-2	No information

Fw 190 Development Batch

Werk-Nr	Engine	Notes
0001	BMW 139	First prototype, Fw 190 V1
0002	BMW 139	Second prototype, Fw 190 V2
0003	(BMW 139)	Third prototype, Fw 190 V3. Used for spares
0004	(BMW 139)	Fourth prototype, Fw 190 V4. Used for static tests
0005	BMW 801C-0	Fw 190 V5 prototype, revised wing, new engine
0006	BMW 801C-0	First Fw 190A-0 development aircraft, as Fw 190 V6, later Fw 190A-0/U1
0007	—	Fw 190A-0 for static tests
0008	BMW 801C-0	Fw 190A-0/U-2, short wing
0009	BMW 801	Fw 190A-0, short wing
0010	BMW 801C-0	Fw 190A-0/U-2, short wing
0011	BMW 801C-0	Fw 190A-0, short wing
0012	BMW 801C-0	Fw 190A-0/U-2, short wing
0013	BMW 801C-0	Fw 190A-0/U-2, short wing
0014	BMW 801D	Fw 190A-0/U-2 short wing. First 801D installation later as Fw 190A-0/U2/U13
0015	BMW 801C-1	Fw 190A-0/U11 (first with long wing) BMW 801C-1 first installation
0016	BMW 801	Fw 190A-0
0017	BMW 801	Fw 190A-0
0018	BMW 801C-1	Fw 190A-0/U5
0019	BMW 801	Fw 190A-0
0020	BMW 801	Fw 190A-0

0021	BMW 801C-0	Fw 190A-0/U3, FuG 25 IFF trials. Crashed 1-10-41
0022	BMW 801C-0	Fw 190A-0/U4. External bombs and fuel tanks. Ejection seat trials later
0023	BMW 801C-0	Fw 190A-0/U4. External bombs and fuel tanks. FuG 16Z radio trials later
0024	BMW 801	Fw 190A-U
0025	BMW 801D	Fw 190A-0/U13, BMW 801D trials
0026	BMW 801D	Fw 190A-0
0027	BMW 801D	Fw 190A-0
0028	BMW 801D	Fw 190A-0/U13, BMW 801D trials
0029	BMW 801	Fw 190A-0
0030	BMW 801C-1	Fw 190A-0/U10
0031	BMW 801D-2	Fw 190A-0/U12. GM1 booster trials
0032	BMW 801	Fw 190A-0
0033	BMW 801	Fw 190A-0
0034	BMW 801	Fw 190A-0
0035	BMW 801	Fw 190 V12=D Series prototype; pressure cabin
0036	DB 603A	Fw 190 V13=D Series prototype
0037	DB 603A	Fw 190 V15=D Series prototype
0038	DB 603A	Fw 190 V16=D Series prototype
0039	Jumo 213	Fw 190 V17, engine test bed, pressure cabin
0040	DB 603A	Fw 190 V18, C Series 'Känguruh'
0041	Jumo 213A	Fw 190 V19=D Series prototype
0042	Jumo 213A	Fw 190 V20=Fw 190D-1 prototype
0043	Jumo 213A	Fw 190 V21=Fw 190D-1 prototype
0044	Jumo 213C	Fw 190 V22=Fw 190D series, pressure cabin
0045	Jumo 213C	Fw 190 V23=Fw 190D series, pressure cabin
0046	BMW 801	Fw 190B-0 series, pressure cabin, long-span wing
0047	BMW 801	Fw 190B-0 series, pressure cabin
0048	BMW 801	Fw 190B-0 series, pressure cabin
0049	BMW 801	Fw 190B-0 series, pressure cabin
0050	Jumo 213A	Fw 190 V25=Fw 190D-1 prototype
0051	Jumo 213A	Fw 190 V26=Fw 190D-2 prototype
0052	Jumo 213A	Fw 190 V27=Fw 190D-2 prototype
0053	Jumo 213A	Fw 190 V28=Fw 190D-1 prototype
0054	Jumo 213E	Fw 190 V29, high-altitude variant
0055	Jumo 213E	Fw 190 V30, high-altitude variant
0056	Jumo 213E	Fw 190 V31, high-altitude variant
0057	Jumo 213E	Fw 190 V32, high-altitude variant
0058	Jumo 213E	Fw 190 V33, high-altitude variant
0059	—	No information
0060	—	No information

Fw 190 Production Variants

Designation	Power Plant	Armament* 1	2	3	Notes
Fw 190A-1	BMW 801C-1	2xMG 17	2xMG 17	2xMG FF	Provision for bomb or fuel tank under fuselage
Fw 190A-2	BMW 801C-2	2xMG 17	2xMG 151	2xMG FF	As A-1 with engine and armament change
Fw 190A-3	BMW 801D-2	2xMG 17	2xMG 151	2xMG FF	As A-2 with engine change
Fw 190A-4	BMW 801D-2	2xMG 17	2xMG 151	2xMG FF	As A-3 with FuG 16Z radio in place of FuG VIIa
Fw 190A-5	BMW 801D-2	2xMG 17	2xMG 151	2xMG FF	As A-4 with lengthened fuselage; EKa 16 camera
Fw 190A-6	BMW 801D-2	2xMG 17	2xMG 151	2xMG 151	As A-5, modified wing, revised armament
Fw 190A-7	BMW 801D-2	2xMG 131	2xMG 151	2xMG 151	As A-6, revised armament, Revi 16b gunsight in place of C 12d
Fw 190A-8	BMW 801D-2	2xMG 131	2xMG 151	2xMG 151	As A-7 with FuG 16Z-Y radio; repositioned fuselage bomb rack
Fw 190D-9	Jumo 213A	2xMG 131	2xMG 151	—	New fuselage, new engine, new wings. Underwing provision for Wfr Gr 21
Fw 190F-1	BMW 801D-2	2xMG 17	2xMG 151	—	Fighter bomber version of Fw 190A-4; fuselage bomb rack plus wing bomb racks
Fw 190F-2	BMW 801D-2	2xMG 17	2xMG 151	—	As F-1, based on A-5, with new clear-view canopy
Fw 190F-3	BMW 801D-2	2xMG 17	2xMG 151	—	As F-1, based on A-6
Fw 190F-8	BMW 801D-2	2xMG 131	2xMG 151	—	As F-1, based on A-8
Fw 190F-9	BMW 801TS	2xMG 131	2xMG 151	—	As F-8 with engine change
Fw 190G-1	BMW 801D-2	2xMG 17	2xMG 151	—	Long-range fighter-bomber, based on A-4, fuselage bomb rack plus underwing fuel tanks
Fw 190G-2	BMW 801D-2	2xMG 17	2xMG 151	—	As G-1, based on A-5
Fw 190G-3	BMW 801D-2	2xMG 17	2xMG 151	—	As G-2 with PKS 11 autopilot
Fw 190G-8	BMW 801D-2	2xMG 131	2xMG 151	—	As G-2, based on A-8

* Position 1, upper front fuselage; position 2, wing roots; position 3, outer wing bays.

Ta 152 Prototypes

Designation	Werk-Nr	Remarks
Ta 152 V1	11 0001	Ta 152H configuration
Ta 152 V2	11 0002	As V1
Ta 152 V3	11 0003	As V1
Ta 152 V4	11 0004	As V1
Ta 152 V5	11 0005	As V1
Ta 152 V6	11 0006	Ta 152C–0 prototype
Ta 152 V7	11 0007	Ta 152C–0 prototype, in /R11 configuration
Ta 152 V8	11 0008	Ta 152C–0/C–1 prototype
Ta 152 V9	11 0009	Intended as Ta 152E prototype. Static tests
Ta 152 V10	11 0010	No information, believed not built
Ta 152 V11	11 0011	As V10
Ta 152 V12	11 0012	As V10
Ta 152 V13	11 0013	Ta 152C–0 prototype, cancelled
Ta 152 V14	11 0014	Intended as Ta 152E/R1. Static tests
Ta 152 V15	11 0015	As V13
Ta 152 V16	11 0016	Ta 152C–3 (C–2) configuration
Ta 152 V17	11 0017	Ta 152C–3 prototype
Ta 152 V18	11 0018	Ta 152C–3 prototype, cancelled
Ta 157 V19	11 0019	Intended as Ta 152B–5/R11. Became Ta 152C–5 prototype
Ta 152 V20	11 0020	As V19
Ta 152 V21	11 0021	As V19
Ta 152 V22	11 0022	Ta 152C–4 prototype cancelled
Ta 152 V23	11 0023	As V22
Ta 152 V24	11 0024	As V22
Ta 152 V25	11 0025	Intended as Ta 152H–2/R11 prototype. Wings used on Fw 190 V32/U2
Ta 152 V26	11 0026	Ta 152H–10 (E–2) prototype
Ta 152 V27	15 0027	Ta 152H–0 modified to Ta 152C–3
Ta 152 V28	15 0030	As V27

Fw 190 Rüstsatz (Field conversion set) Numbers

Identity	Equipment	Application
R1	FuG 16Z–E communications radio for lead fighters	Fw 190A–4, A–5, B–1
R1	WB 151/20 containing two 20-mm MG 151/20E guns beneath each wing, with 125 rpg	Fw 190A–6, A–7, A–8, A–9, A–10, F–3, F–8, F–15, G–3, D–9, D–12
R2	One 30-mm MK 108 gun beneath each wing	Fw 190A–6, A–7, A–8, A–9, A–10, F–8, D–12
R3	One 30-mm MK 103 gun beneath each wing	Fw 190A–6, A–7, A–8, A–9, A–10, F–3, F–8
R4	GM 1 nitrous oxide power boost	Fw 190A–6, A–8, G–8
R5	Long-range fuel tank installation, initially comprising one 25 Imp gal (115 litre) tank in fuselage	Fw 190A–8, F–8, F–16, G–3, G–8
	Subsequently, additional fuel tanks in wing – four tanks of 69 Imp gal (315 litre) total capacity	Fw 190D–12, D–13
R6	Two Wfr Gr 21 rocket projectile launchers beneath each wing	Fw 190A–4, A–5, A–6, A–7
R7	Additional armour plate protection	Fw 190A–8
R8	Additional armour plus one 30-mm MK 108 gun in each wing outer bay	Fw 190A–8, A–9
R11	All-weather equipment comprising heated cabin windows, FuG 125 radio and PKS 12 auto-pilot	Fw 190A–8, A–9, D–9, D–11, D–12, D–13, D–15, Ta–152, B–5, C–0, C–1, C–2, C–3, H–0, H–1
R12	As R11 plus one 30-mm MK 108 gun in each outer wing bay	Fw 190A–8, A–9
R13	Night-fighter equipment comprising FuG 16ZS and FuG 125 radio, flame damping on exhausts, etc	Fw 190F–8, F–9
R14	ETC 502 torpedo rack, lengthened tailwheel and fuselage stiffening	Fw 190F–8, F–9, D–9, Ta 152C–1
R15	ETC 502 to carry BT 1400, TSA 11A sight, PKS 12, etc	Fw 190F–8, F–9
R16	As R15 for BT 700	Fw 190F–8, F–9
R20	High pressure MW (methanol-water) tanks in fuselage	Fw 190D–11, D–12, D–13
R21	Combination of R11 and R20	Fw 190D–9, D–11, D–12, D–13, Ta 152H–1
R25	Jumo 213EB plus R5 plus R11 and high-pressure MW tank in port wing	Fw 190D–12
R31	Combination of R21 and R4	Ta 152C–1, Ta 152H–1